SAFETY INCENTIVES

THE CHRONICLE-HERALD

MINING TRAGEDY

11 bodies pulled from Westray pit; rescuers hunt for 15 men

[illegible]

[illegible]

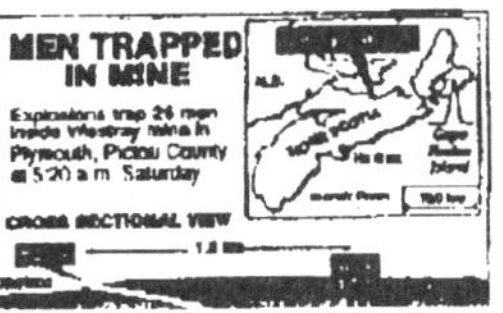

Families' dread turns to anguish

The Westray coal mine accident killed 26 miners in Nova Scotia, Canada in 1992. One month earlier, the mine had received a major safety award. An investigation revealed that the mine's production bonus system was not conducive to safety.

SAFETY INCENTIVES

The Pros and Cons of Award and Recognition Programs

Wayne G. Pardy

MORAN ASSOCIATES
Orange Park, FL • Washington, DC

MORAN ASSOCIATES
1600 Brighton Bluff Court
Orange Park, FL 32073-7409

Phone: 904-278-5155
Toll-Free: 1-800-597-2040
Fax: 904-278-5494
E-mail: sales@moranbooks.com
Web Site: http://www.moranbooks.com

First Edition ISBN 1-890966-53-3

Printed in the United States of America

03 02 01 00 99 5 4 3 2 1

Library of Congress Catalog Card Number: 99-61521

This book is dedicated to the past, present and future.

To the past:

For my father, Garland, whose achievements were hard fought but rightfully earned.

To the present:

For my wife, Lisa, my mother, Jean, and to Sheila and Bill, for all the encouragement and support. This is as much your book as it is mine.

To the future:

For my children, Beth and Mark: may you always remember the achievements of the past, enjoy today, and strive for the achievements of tomorrow.

TABLE OF CONTENTS

FOREWORD xi

PREFACE xv

ACKNOWLEDGEMENTS xxi

ABOUT THE AUTHOR xxiii

CHAPTER 1: WHAT ARE INCENTIVES? 3
Incentives: Popular or Detested? 11
Compliance and Control: Practical Incentive Strategies? 18
An Example of Safety Incentives Gone Wrong 19

CHAPTER 2: SAFETY MANAGEMENT AND INCENTIVE STRATEGIES, PAST AND PRESENT 23
Incentives for Compliance 28
You Get What You Pay For 31
Strategic Planning: A Safety Management Imperative 36
Rewarding Achievement-Based Safety 36
Risk and Reward: A Risk Management Approach 41

CHAPTER 3: MOTIVATION AND LEADERSHIP 47
A Shared Approach with Common Goals 49
The Internal Responsibility System for Occupational Health and Safety 50
Destroy the Safety Myths and Cliches 52
Theory and Reality: What are the Causes of Accidents and Injuries? 60
Leadership: A Safety Management Imperative 64

CHAPTER 4: WHERE DO SAFETY INCENTIVE AND RECOGNITION PROGRAMS FIT IN? 75
The Safety/Quality Relationship 75
Caution and Intelligence Needed with Selected Quality Approaches 81
Avoiding Dysfunction: Do Your Homework 84
A Tradition Continues, But at What Price? 87
Other Real World Applications 89
More Sophisticated Approach Needed 94

CHAPTER FIVE: DO INCENTIVES WORK OR NOT? 97
Labor vs. Behavior-Based Safety: No Love Lost 100
Beware of Dysfunction 107
Building a Solid Foundation 110
What Does the Future Hold? 113

CHAPTER 6: CULTURES AND EMPOWERMENT 115
Culture is Important, But Keep Your Perspective 115
Whose Behavior: Management, Workers or Both? 116
Keys to Cultural Success 117
Culture in the Real World 119
Beware the Steel Hand in the Velvet Glove 122
Getting Beyond the Empowerment Rhetoric: Meaningful Incentive and Recognition Considerations 124
Brother, Can You Spare a Paradigm? 126
Conventional Safety Strategies and Structures: Fast Becoming Obsolete? 129
Adapt or Else: Flexibility is the Key 130
Safety Can Only Happen Through People 130
Empowerment Needs Attention to Detail 133
Welcome to the Real World! 135

CHAPTER 7: SAFETY PERFORMANCE MEASUREMENT 137
Measuring Process Performance 138

Safety Performance Measurement Options for Strategic Planning and Recognition 144

CHAPTER 8: EXPERIENCES WITH INCENTIVE AND RECOGNITION PROGRAMS 155
Do They Work? 155
Or Don't They? 159
A Veteran's Perspective 163
Success Factors 166
Should You Have an Incentive or Recognition Program? 170

CHAPTER 9: EMPLOYER INCENTIVES FOR SAFETY AND COMPLIANCE 173
More Balance Needed 174
Benefits of Positive Incentives for Employer Compliance 177
Voluntary Protection Options 178

CHAPTER 10: THE FUTURE OF SAFETY AND INCENTIVE STRATEGIES 181
X vs. Y 184
High Performance Recognition and Incentives 186
Why Does the Tail Wag the Dog? 187
Collaboration vs. Compliance 188
Conclusion 189

APPENDIX A: SAFETY INCENTIVE, RECOGNITION AND AWARD SYSTEM PLANNING 191

APPENDIX B: SAMPLE SAFETY RECOGNITION AWARD PROGRAM 195

APPENDIX C: WORKPLACE HEALTH, SAFETY & COMPENSATION COMMISSION OF NEWFOUNDLAND AND LABRADOR LEADERSHIP RECOGNITION AWARDS

PROGRAM APPLICATION FORMS 203

APPENDIX D: OSHA VOLUNTARY PROTECTION PROGRAM INFORMATION AND APPLICATION FORMS 211

APPENDIX E: SAFETY INCENTIVE PRODUCTS AND SERVICES 235

NOTES 253

INDEX 267

FOREWORD

There is no more hotly-debated issue in safety than safety incentive programs. Are they good, bad, harmful to safety, flawed, or just bad practice? This highly-emotional subject has been at the top of the list of hot topics for well over 20 years.

In reality, the roots of this debate lie in the fact that in the beginning, neither management nor those who practiced safety knew-how to construct an effective and comprehensive safety and health program for workers. Incentives followed a well-established management paradigm: if you can't manage it, throw money at it.

This may seem like a pessimistic judgement of safety incentive programs. Not so. Looking at the long-term and widely-established use of incentives as a primary method for injury reduction, no one can dismiss this practice as totally ineffective or bad safety practice. It would be like casting aside sales commissions as being a disincentive to sales.

Still, many would eradicate the practice of offering incentives for safety, due to undeniable problems incentives can cause, such as underreporting of injuries. The fact remains that there is poignant and undeniable evidence of the long-term effectiveness of safety incentives.

The problems associated with safety incentives do not lie in the basic concept, only in the design and implementation of individual incentive programs or applications. Those who argue for the complete elimination of safety incentives don't have the facts to back them up. In reality, the debate over the efficacy of safety incentives is moot. But incentives continue to be the most hotly-debated and most widely-discussed topic in safety publications today.

Accepting the overall efficacy of incentive programs by their historical success, the question that safety practitioners should naturally ask is, "What aspects of safety incentive programs work and why?" Many authors have provided ideas that attempt to answer parts of this question. These efforts have been only marginally effective, though, in that they have only dealt with selected aspects of incentive programs, have been poorly researched, have only discussed incentives as part of the coverage of another subject (such as behavior-based safety), or have been emotional rantings.

The lack of a well-researched and documented discussion of both the global aspects and specific applications of safety incentive programs has been sorely lacking. Until now.

Using a methodical approach, Wayne Pardy has provided the safety community with what will undoubtedly become the authoritative guide to safety incentive programs. Mixing theory and practice, Pardy provides a clear formula for success. But one of the major strengths of this book is that it brings together successful concepts and strategies from outside the traditional practice of safety and, using them as support and wisdom, bringing clarity to this highly-debated subject.

In my studies of paradigm shifts, I have frequently found that new ideas frequently come from sources outside our own backyards. So it comes as no surprise to me that it took a Canadian from Newfoundland to bring fresh ideas and much-needed documentation to such an important subject. This book will not only free the concept of safety incentives from the pall of suspicion, I am confident that it will redefine and standardize the applications of these still-effective safety tools.

F. David Pierce
President
Society for the Advancement of Safety and Health

PREFACE

Safety incentives, recognition and award programs are some of the more controversial, topical and divisive issues in industrial safety. No matter who you speak with, they will all have an opinion, success story or horror story associated with safety incentives, and they are all too quick to pass along their opinion as representative of the current thinking on safety recognition and incentive programs, safety management and human behavior.

Many safety professionals have discovered that when safety programs and incentives are used in the same sentence, questions arise about the focus, logic, rationale and emphasis of their industrial safety efforts. The incentive and recognition debate may be such a flash point because the key focus of these approaches deals with people and their behaviors, attitudes and actions.

Consider how we recognize and reward people for safety achievements, or attempt to otherwise motivate workers to work safely. Traditional approaches have held that if you offer some prize to workers, and workers, either collectively or individually, are fortunate enough to work for a period of time without sustaining any disabling injuries, then those workers will be safer. Many in labor hold the belief that these awards, trinkets, or other 'carrots' are nothing more than another way in which management attempts to influence the behavior of workers.

Many in management believe that workers need to be motivated to work safely, and by offering incentives, workers will perform as desired. It's no surprise then that management and labor hold entirely different views about the use and effectiveness of safety incentive and recognition programs.

It's interesting that safety legislation, in practically in every part of the world, has as its focus the physical work environment. Yet some would argue that most accidents are the result of action or lack of action by employees and as a result, rules, regulations and standards cannot be the driving force behind safe and efficient operations.

The research for this book began after I presented a paper and lecture entitled "Safety Incentive, Recognition and Awareness Programs: One Company's Experience and an Industry Perspective" in Toronto, Canada, in February 1997. The promotional brochure for the conference contained an interesting statistic: 90% of all accidents are caused by unsafe behavior. The point being pushed by the conference sponsors was that if you could identify, track and monitor unsafe worker behavior, you could prevent accidents.

One of the conference's seminars was devoted to behavior-based safety. Out of curiosity, I called the individual listed as that session's presenter and asked where this 90% figure came from. She indicated that it had been provided by consultants who supply training on the behavior-based approach to workplace safety management. I asked where these consultants got their information, and she indicated it was from the many employers with whom these consultants discussed ideas on accident causation.

Having talked to many employers and employer representatives myself, it was not surprising that many of them feel that worker behavior is the key to both accident causation and prevention. But nowhere in my research and discussions with safety practitioners was I able to verify any statistically valid studies about the long-term impact of employee behavior modification approaches on a total workplace safety management system. Short term compliance and obedience, yes.

The concept of human behavior as one of the main causes of accidents, juxtaposed with the focus and assertion of government regulators that the physical conditions of the workplace need to be the key employer focus, is an interesting paradox. Other studies, particularly those coming from the TQM arena, suggest that over 80% of all problems in many companies (we can assume, for the sake of argument, that accidents and injuries can be defined as a problem) can be solved by management, and that accidents and injuries sustained by workers are merely symptomatic of other organizational problems, or problems with what many safety practitioners are now calling the management system.

To simply say that the unsafe acts of workers are the cause of safety problems gives short shrift to decades of work and effort on the part of business, labor and government, with particular emphasis on establishing standards for safe workplaces. For many safety professionals, the challenge to both comply with tougher and more stringent legislation, while at the same time trying to come to grips with what some have deemed the people side of safety, can lead to frustration.

Perhaps this frustration eventually creates a type of safety schizophrenia in those with responsibility for accident prevention, forcing them to be motivated in their accident prevention efforts by two seemingly contradictory or conflicting principles: legislative safety standards and efficient, effective production. Such is the lot of today's safety professional.

Fundamental to successful safety performance is the degree to which a business and its resources are deployed to deal with preventing accidents, occupational illnesses, and other types of accidents which continue to cost society untold millions of dollars in lost wages, productivity, efficiency, legal costs, human suffering, and regulatory enforcement initiatives.

But how can business, labor and government harness the knowledge and skill of its work force to address and eventually solve many of the traditional safety problems which continue to plague industry? This can be accomplished through the power of all people having a vested interest in successful safety performance.

Simple worker behavior manipulation or modification schemes may not be able to achieve this strategic, long-term success. Approaches to modern safety management which address the entire safety system are needed. By the same token, if behaviors are to be the focus, all worker and managerial behaviors need to be strategically addressed.

A total quality management approach to health and safety management tends to offer more long-lasting results, while behavior management generates a quicker impact. In general, a total quality management approach to safety has its focus on the entire system, which in turn has the ability to result in deeper cultural changes within the organization, including the management system.[1]

This book provides a thorough look at the subject of safety incentives, where they've come from, how and why they are used, why some people detest them, and why some swear by them.

In addition to exploring traditional and novel safety incentive approaches, it will suggest ways in which workers and management can come together to develop a common strategy, an achievement-based safety culture. And it will explore the proposition that if incentives and recognition programs help workers improve their safety performance, why would not the same logic apply for safety incentives for government compliance options, or safety incentives at the middle and senior management level?

An achievement-based safety culture does not exist for the exclusive benefit of one group at the expense of another. Whereas traditional safety incentive and recognition systems mainly center on worker behavior and the reduction or elimination of unsafe acts, an achievement-based safety culture demands superior performance from all levels of the business. Cascading responsibilities result in cascading performance standards and expectations, and provide top to bottom synergy for all levels of a business. Of course, the three conventional E's of safety—Engineering, Education, Enforcement—continue to be the major focus of many industrial safety efforts, and indeed they have had, and will continue to have, an important role to play.

We've seen all too much of the potentially harmful effects of an injury or accident-based safety culture. That focus is narrow, restricted, and misdirected. As we prepare to enter the 21st century, we need to raise our expectations for industrial safety performance. This is where an achievement-based safety culture can be invaluable. It's about short and long-term safety performance as opposed to immediate rule compliance. It's about collaboration versus simple compliance.

You cannot force people to cooperate or collaborate, to work together, or to simply follow safety rules. Memos from the president or CEO on the value of teamwork may be great prose, but may do nothing to affect the realities of the shop floor. Some would even go so far as to say that while the memos are written for those on the shop floor, the CEO has never even been on the shop floor.

When looking at safety incentives and recognition, keep in mind the results of a recent survey of workers and management. It found poor upward communication and a lack of trust, with general employee dissatisfaction with recognition and incentive programs defining this gap. In fact, the biggest divide was in the area of performance management and recognition. While 57% of senior management felt their businesses rewarded good performance, only 20% of employees agreed.

This book is not meant to be a blueprint for handling safety incentives, improving safety performance or modifying safety cultures. It is not a cookbook, nor does it describe a simple exercise that will make your safety concerns disappear. It is meant to inspire you to challenge and discuss, debate and explore. It is meant to help shed the fluff which sometimes dominates our feel good safety approaches these days, and it cuts to the chase. It is not a silver bullet or a magic cure, but it is a thought-provoking examination of safety, safety cultures and safety incentives. I trust you will find it interesting, and maybe even helpful.

And for those of you who accept the challenge of moving from an injury-based culture to an achievement-based safety culture, enjoy the ride!

Wayne G. Pardy

ACKNOWLEDGEMENTS

As with any undertaking of this magnitude, there are many people who, in one way or another, have influenced both the content and focus of this book. In particular, I'd like to thank Alex Padro of Padro Communications for his confidence in the idea, and his guidance and support throughout the writing and editing process. His assistance and advice have been invaluable.

I'd also like to thank the staff of Moran Associates, who also felt there was both a need and a practical application for the ideas and information presented between these covers. I trust we have been able to help our readers look at this subject in a whole new light.

Thanks also to the countless safety professionals, practitioners, critics and supporters who have expressed a wide variety of views and opinions on this topic. A quick scan of the index will demonstrate that many people have influenced this book. Thanks in particular to Noel Bishop, my colleague in safety for some fifteen years, and with whom I've debated, discussed, agreed and disagreed on just about every issue discussed in this book. To Alfie Kohn, who offered some very cogent arguments against incentives, and provided the basis for the very necessary balance that I've tried to achieve here. To Bill Gushue, who kept saying, "You have to write the book." How right you were. Thanks for the push.

Of course, no list of acknowledgements would be complete without recognizing those who, from time-to-time, must have felt as if it was not so much that a book was being written, but that someone in the basement refused to come out. Without the support and encouragement of my wife Lisa, and my children, Mark and Beth, the labor would have been much harder, and the outcome less satisfying. Additional thanks to Lisa for her critical eye and keen sense for accurate spelling. You can now put away your red pen... until the next one!

ABOUT THE AUTHOR

Wayne G. Pardy is executive consultant of Pardy & Associates, in St. John's, Newfoundland, and the president of Creative Business Solutions Inc., a software company specializing in safety system applications, including audit software, investigation systems and risk management software. For the past 14 years he has also been a safety professional with Newfoundland Power.

Pardy is a graduate of Memorial University of Newfoundland, an internationally Certified Trainer (CT) and a Canadian Registered Safety Professional (CRSP). He has been a frequent speaker at national conferences and professional development seminars, and his work has been featured in a number of international conferences in both the US and Australia. Pardy has also conducted training and consulting for business, government and labor groups throughout Canada.

In addition to his role as advisory board member for the Safety Engineering Technology Program at Cabot College, Pardy has also guest lectured on safety for the School of Continuing Studies and Professional Development, Memorial University of Newfoundland.

Pardy has written extensively on safety, quality and human resource development in numerous American and Canadian publications. He is a past chair of the Editorial Advisory Board for *OHS Canada* magazine, and is currently the contributing editor for safety management for *Canadian Occupational Safety* magazine. He received the 1991 CSSE President's Award for his two part series on "Building an Effective Safety System: Key Strategies for Maximizing Resources and Minimizing Losses" in *Canadian Occupational Safety*. As a contributing editor for

Industrial Safety & Hygiene News, his feature article on "Worker Empowerment and Joint Health and Safety Committees" won the 1994 Chilton Editorial Award. This was the first time this award was won by a Canadian.

Pardy lives in St. John's, Newfoundland, with his wife, Lisa, and two children, Beth and Mark.

SAFETY INCENTIVES

CHAPTER 1

WHAT ARE INCENTIVES?

No matter who you ask, almost everyone agrees that good safety performance needs to be recognized and rewarded. On that key point, there is general agreement. But while safety incentive and recognition awards have been around for some time, there is very little agreement on their value, and upon which principles to base their implementation or eradication.

Despite theoretical evidence, behavioral modification approaches, common sense and anecdotal approaches, humanistic approaches, philosophical arguments and empirical research, opinions on safety recognition and incentive programs still vary, and vary dramatically.

Scratch a safety professional and you'll find an opinion on safety incentives underneath. The difficult part is coming up with consistent principles, definitions or guidelines upon which to base safety incentive and recognition approaches, how they can add value to existing safety systems, and help promote outstanding safety performance. Still greater difficulty exists in proving these programs actually result in definitive performance improvement over the long term. In other words, can we prove a definite cause and effect relationship, and is that relationship desirable, strategic and reasonable?

If a recognition or reward system is set up, and injury rates go down, does this necessarily mean that the program was the direct cause, or were there other factors? Conversely, if your injury rates and claims costs go through the roof, in spite of the fact that you may have a very comprehensive safety incentive scheme in place, what would you do next?

Safety performance recognition has existed under a number of different names for quite some time, but if we hope to use these tools and techniques correctly, we've got to learn whether, when and how best to use them. If used, they must be used with insight, intelligence, honesty, integrity and with consideration for both the minds and bodies of those they are intended to recognize or motivate, and not simply as another extension of the traditional command and control model. In *The Service Edge*, authors Ron Zemke and Dick Schaaf write, "Incentive systems aren't automatic performance generators. ... They can even backfire and be counterproductive when they don't work out in the fashion intended." [1]

In a survey published in the May/June 1997 edition of *Canadian Occupational Safety*, respondents were almost equally split over their safety incentive beliefs and practices. By far, the most common incentive programs used by those who responded to the survey were those based on being accident free for a specific period of time.

Many of the programs profiled consisted of incremental stages of year-long accident-free periods (i.e. one year, two years, five years, etc.), while others tied the incentives to reducing injury statistics from the previous year.[2] 2 It is these types of accident-free incentives which have drawn the most criticism, especially from government and labor. Yet, they still appear to be the most popular, while at the same time being the most misused, maligned and misunderstood.

Perhaps the best place to start is to examine dictionary definitions and see if the words incentive, recognition and award actually mean what we think they mean. Generally speaking, incentives are rewards with some strings attached, commonly referred to as the carrot and stick approach. The presumption is that if you do certain things or reach certain goals, you will receive your reward.

incentive: n. 1. A motivating influence, stimulus. 2. a. An additional payment made to employees as a means of increasing production. b. (as modifier): an incentive scheme. adj. 3. Serving to incite to action.[3]

The rewards associated with incentives are usually financial in nature, or hold some other monetary value. Pay for performance, incentive compensation and commission sales schemes, for example, are included under these incentives programs. In *The Reengineering Revolution*, Hammer and Stanton define incentives as "inducements, both positive and negative, to get people to behave as required by the reengineering effort. They include, but are not limited to, financial incentives. Continued employment is also a powerful incentive, and one that can be effective. Confronting resisters and making it clear that termination will be the consequence of their behavior is a very valid technique. But there are other forms of incentive as well: sharing in the psychic rewards from a successful effort, having new career opportunities, working in a more fulfilling job." [4]

recognition: 1. The act of recognizing or fact of being recognized. 2. Acceptance of acknowledgement of a claim, duty, fact, truth, etc. 3. a. A token of thanks or acknowledgement.[5]

Though somewhat similar, the motivational power in recognition lies mainly in its ability to appeal to an employee's sense of pride. It's the 'pat on the back,' the 'coffee and donuts with the CEO,' or the 'congratulations on a job well-done' type of system. The important thing to remember about recognition is that different people like to be recognized differently. One person's plaque or trophy proudly displayed in the recreation room may be another person's basement junk. Essentially, recognition is a reward that costs less than incentives. Find out what makes a person tick and you can use recognition. At least, that's the theory.

award: 1. To give (something due) esp. as a reward for merit: to award prizes. ... Something awarded, such as a prize or medal: an award for bravery.[6]

Awards can be given for an achievement, milestone, accomplishment or behavior that a business deems important or worthy of note or celebration. They can involve small gifts, plaques, memos, or any other type of award which the winners and presenters of the award feel will motivate as required. They are meant to acknowledge a job well done and motivate others to behave likewise. Motivation and behavior are key words that get used with great regularly in the safety incentive debate.

Safety professionals and others with an interest in safety have for many years been trying to adapt various behavioral and motivational approaches to safety in order to get to the root causes of accidents and injuries in their respective businesses. But this approach hasn't been without controversy. While admittedly honorable in their intent, many safety incentive approaches have produced the exact opposite of what they were intended to achieve. As a result of having been used rather haphazardly, representatives of organized labor and others have come to view them as just another management manipulation scheme, with undue emphasis on the worker rather than on the workplace, the people who control the workplace, and the means of production.

Just how controversial traditional worker safety incentives can be is reflected by the October 8, 1996 comments of the president of the Canadian Labor Congress at a conference on disability and work. Bob White stated that on-the-job hazards are costing billions, yet some employers are bribing workers to keep work injuries secret. Employers should be making their operations safer, not running contests that tempt workers to lie, said White. "Many companies run programs designed to reduce reporting of workplace injuries by offering prizes, like company jackets or even vacations abroad, to the worker or team that goes the longest time without filing a compensation claim," White said. "Why not concentrate on the real source of the costs by reducing the injury rates, deaths and diseases that are caused by the workplace?"[7]

In the September 28, 1989 issue of *ENR* magazine, one company was shown to be offering pizza, savings bonds and jackets to employees and groups of employees who attained the often elusive, sometimes manipulative, and often artificial distinction of being lost-time-injury-free for one year. Some of the employee comments reflected in the article included:

"I was so frightened that I was going to spoil the record of 400 days that you can't imagine."

"Nobody wants to be the one to spoil the record."

"I think that if somebody spoils the record or gets hurt, he'll get buried in concrete so nobody will ever find out about it and they won't lose the record."[8]

No doubt these comments were perhaps made in a half joking manner, but they also reflect how the safety award system at that company operated. Many feel that by using safety incentives, peer pressure can be applied which, in the opinion of safety incentive advocates, helps improve safety performance. In reality, those employees' comments may be symptomatic of the serious dysfunction of many safety incentive approaches used in North America, and indeed entire safety philosophies and resulting safety management systems worldwide.

In the opinion of Dr. Robert Sass in a 1984 paper entitled, "*The Value of Safety Contests: A Point of View,*" "The tacit assumption behind safety contests is that workers are primarily responsible for accidents through their carelessness, accident proneness or bad attitudes, and that they can therefore stop accidents from happening merely by resolving to be more careful, to obey their supervisors or to have a positive attitude towards safety. ... Carelessness and accident proneness are false accident causation theories which blame the victim, the worker, for accidents and deflect attention from unsafe conditions." [9]

Perhaps the saddest and most dysfunctional example of a safety system was reflected in the Westray Mine Public Inquiry Report. In the early morning of May 9, 1992, a violent explosion ripped through the depths of the Westray coal mine in Pictou County, Nova Scotia, Canada, killing all 26 miners working there at the time. While many terrible stories were told of the events leading up to that explosion, interesting information came to light about safety at the mine, or perhaps more accurately, the lack of safety there. Even more interesting was the evidence presented about the use of incentives at the mine, and the safety award given to the mine just one month before the fatal explosion.

The John T. Ryan Trophy was first introduced in 1941 by the Mine Safety Appliances Company of Canada (now MSA Canada Inc.) to promote mine safety achievements. It is a much-sought-after award in the mining industry and eligibility for the award is based on the frequency of reportable injuries. While Westray management was notorious for its ignorance of safety, it apparently was interested in the outward trappings of safety. At the suggestion of the managing director of the Chamber of Mineral Resources in Nova Scotia, Westray applied for the John T. Ryan Trophy, and won. In an April 9, 1992 memo to employees, the vice president and general manager of the Westray mine congratulated the workers on their achievement.[10]

The problem with the award, and other safety awards of this type is that they say nothing about safety or risk management. They simply tell you that for a given period, usually the artificial period of one year, lost time injury frequency or severity was low, but it doesn't explain why. How could a mine which was proven to be so inherently unsafe, as evidenced by the lack of any semblance of a safety culture, and which eventually killed 26 miners, receive such a safety award, only a month or so prior to a fatal explosion?

It also shows the folly of a third party accepting an award nomination without any independent audit or verification of the validity of the integrity of the nomination. Later in this book, we'll see how a high performance or achievement-based safety incentive and recognition systems can improve safety performance, especially when compared with low performance, injury or statistic-based systems.

A key finding of the Westray report was the issue of the production bonus system. While the Ryan Trophy was an example of an award which some, for whatever reason, felt was important to acquire, the production bonus at Westray was a good example of a variation of incentive pay schemes. In the evidence presented at the Westray inquiry, the production bonus involved different bonus percentages, and depending on the work being done, workers could earn a bonus for average monthly production in excess of 500 tons per machine shift. Workers had a percentage of their bonus deducted if they missed work: a one day absence meant a 25% reduction and a two day absence meant a 50% reduction.

The finding of the inquiry concluded that it was clear from the evidence presented by the miners, and from an outside expert's analysis of that evidence, that the incentive bonus scheme based solely on productivity was not conducive to safety at Westray. Indeed, Commissioner K. Peter Richard recommended that incentive bonuses based solely on productivity have no place in a hazardous working environment such as an underground coal mine.

In the opinion of the Commission, these types of incentive systems should be replaced, if practical, with a safety incentive system which incorporates the principles of shared, cooperative development between workers and management and rewards for both individual and group performance for all workers in the organization. It was felt by the Commissioner, based on the evidence and research presented at the inquiry, that by following these guiding principles, safety incentive plans can enhance productivity.[11] These incentives, if structured with intelligence, integrity and with a total safety management approach, can benefit workers, businesses, governments and society in general.

While some may consider the issue of safety incentives to be trivial, this approach to managing workplace health and safety has become so widely accepted as to seriously jeopardize many fine and worthwhile efforts which government, labor and business have struggled to initiate and implement over the past 50 years. More than that, the safety incentive debate reflects the views and focus of many other aspects of safety philosophy and approaches to safety management in general. They are part and parcel of any intelligent approach to the many varied and complex issues that face all businesses in attempting to structure an effective safety management system.

INCENTIVES: POPULAR OR DETESTED?

There are as many different opinions on the value and effectiveness of incentive schemes as there are people who use them. Different individuals and businesses use them for different reasons, and have various degrees of knowledge and understanding of incentive theory and practice. There's no doubt they receive great attention, but do they work? And what are our criteria for deciding whether they work?

Aside from how important it is to recognize good safety performance, and presumably motivate people to work safely, many individuals, when asked to rationalize the logic of their incentive approaches and the expectations they have of their approaches, find themselves at a loss for words. Still more are further silenced when are asked if there is a direct cause and effect relationship between the use of safety incentives and statistical safety performance in the form of reduced accident and injury frequency or severity.

Many will suggest that after they introduced safety incentives, statistics improved 50% or 80%. But what many have difficulty determining is whether there is a direct cause and effect relationship, or whether the improvement in safety performance and the introduction of a safety incentive approach were merely coincidental.

Many studies have produced anecdotal and empirical evidence suggesting that the use of safety incentives does indeed result in reduced accident and injury frequency and severity. There have also been many studies and popular articles published which support the notion that rewards and incentives are nothing more than simple attempts to control and manipulate the behavior of others, and that they simply do nothing more than promote a command and control type of compliance, and force the subjects to do nothing more than exhibit the correct behaviors in order to receive the prize, whatever that prize may be.

It may be that the number of supposed success stories and articles on safety incentives are greater because the use of these approaches is perceived to be very positive. Why would wanting to provide incentives and recognition be a negative thing? Surely if you're against safety incentives, you have a negative attitude towards safety. Taking a page from the behaviorist's book, my attitude is very positive, but my behavior towards safety incentives, as reflected by my words between the covers of this book, may sometimes be perceived as negative, and in need of attention.

Positive incentives can create new opportunities for safety management and improved levels of safety performance. But they have to be used with intelligence, integrity, honesty, and with the best interests of all workers, management and the long-term viability of the business in mind.

There are as many different approaches to safety incentives and recognition options as there are organized health and safety management systems. The challenge is creating an approach that meets existing business and regulatory realities and is tied to strategic safety goals, objectives and targets.

In a 1978 edition of the *Journal of Applied Psychology*, Judi Komaki, Kenneth Barwick and Lawrence Scott of the Georgia Institute of Technology published a study entitled, "A Behavioral Approach to Occupational Safety: Pinpointing And Reinforcing Safe Performance in a Food Manufacturing Plant." Using a behavioral approach, the experiment conducted a baseline analysis which discovered that safe practices were probably not being maintained because there was little, if any, positive reinforcement for performing safely. At the same time, employees were not provided with opportunities to learn to avoid unsafe practices. When employees did perform unsafely, they rarely, compared to the number of unsafe acts, experienced an injury. Using a behavioral observation approach that specified safe, unsafe or unobserved behaviors, scores were computed which translated into levels or degrees of safety. In addition to training on the behavioral observation techniques, and a departmental goal of 90% safe that was suggested and agreed to by employees, data began to be collected and progress and targets charted.

To complement the feedback on how many safe or unsafe behaviors were recorded, supervisors were to also recognize workers when they performed select activities safely. One of the fundamental issues for this approach was to clearly define the expectations for safety in terms of the behaviors of the workers who were performing the work. By specifying very exact behaviors and defining them as being safe, the objective assessment was meant to make safety an objective, observable item. Rather than using slogans and posters which extolled the virtue and value of general safety sentiments, safety was defined as very specific, observable behaviors. The research noted that although a correlational analysis was not possible for the study, it was hoped that assessments of the relationship between the behavioral measure of safety and injuries would be conducted in future studies.[12]

The long term effects of a 'token economy' on safety performance in open pit mining were studied by David K. Fox, B.L. Hopkins and W. Kent Anger, with the results published in the Fall 1987 edition of the *Journal of Applied Behavior Analysis*. The authors noted that most of the research conducted using the behavioral approach to safety focused on changing behaviors, or behavior-produced environmental conditions assumed to be unsafe. In attempting to determine a direct cause and effect relationship between behavioral approaches to safety and improved safety performance, the authors noted that the research did not examine changes in the number of incidents, or the extent of injuries as the behavioral changes occurred.

In the Fox, Hopkins, Anger study, trading stamps were used, whereby employees earned stamps for working without a lost time injury, for working in a group in which other workers experienced no lost time injuries, for not being involved in equipment-damaging accidents, for making adopted safety suggestions, and for unusual behavior which prevented an injury or accident. Workers lost earned stamps if they or other workers in their group were injured, caused equipment damage, or failed to report accidents or injuries. Stamps could be exchanged for thousands of different items at redemption stores. According to the study, the implementation of the token economy was followed by large reductions in the number of days lost from work because of injuries, the number of lost time injuries and the costs of accidents and injuries, and these improvements were maintained over a period of years.[13]

The effectiveness of various positive approaches to enhance safety performance has supposedly been clearly established by research.[14] In "The Use of Incentives/Feedback to Enhance Work Place Safety: A Critique of the Literature", R. Bruce McAfee and Ashley R. Winn examined some 24 studies which looked at the effectiveness of positive reinforcement and feedback. While one of the more striking findings of the review of the research was that there were so many different forms of rewards and feedback studied, the consistent variable was that most of the approaches were based on the behavior modification principle that behavior is a function of its consequences, and therefore, rewarded behavior is likely to be repeated.[15]

But the long-term implications of these approaches have yet to be clearly established, and some current practices and approaches leave serious questions unanswered. None of the studies considered the long-term ramifications of safety enhancement and accident reduction, and the effects of various independent variables on employee job satisfaction or on the supervisor-subordinate relationship were largely ignored.[16]

It's also interesting to note that many of the studies are very clinical in their assessment of various incentive approaches, and we seldom hear from the subjects on their perceptions of the incentive approach. Nor do we get any understanding of the type of safety climate or culture within which these approaches were used, nor any objective examination of existing safety management systems employed in the respective businesses. Further, there's no evidence to suggest that any other mid to senior management level incentives were used to complement or support the employee-focused safety incentive approaches.

While there has been a considerable amount written on the positive aspects of incentives, there has also been substantial examination given to alternatives to the incentive and recognition approach. The difficulty in examining the positive incentive articles seems to be that they are written in a non-critical way, leaving the reader to accept from the outset that safety incentives are, by definition, beneficial, and to suggest otherwise is just not acceptable.

In the opinion of Alfie Kohn, perhaps one of the best-known of the behaviorism critics, the "token economy" utilized in some safety studies and other behavior modification approaches is nothing more than an elaborate behavior modification plan, with markers or treats which can be traded in for gifts, given out in doggie biscuit fashion when people act appropriately.

According to Kohn's research, when the token reinforcement is removed, desirable responses generally return to baseline or near baseline levels of performance. In everyday language, when the goodies stop, people go right back to acting the way they did before the program began.[17] It's a "do this and you'll get that" approach, and according to Kohn, this idea has become so pervasive and apparently so widely accepted as fundamental to the running of business today that it's time to worry. In Kohn's opinion, "The time to worry is when the idea is so widely shared that we no longer even notice it, when it is so deeply rooted that it feels to us like plain common sense."[18]

Does this mean that safety incentives and reward schemes are not effective? Not exactly, because according to Kohn and others, rewards and punishments do work by inducing compliance. As Kohn notes, if your objective is to get people to obey orders and to do what they're told, then bribing or threatening them may be a sensible strategy.

Kohn's research also notes that reinforcements based on behavioral modification approaches do not generally alter attitudes and emotional commitments. As a result, long-term success or improvement is never realized. In order for a business or institution to exist and thrive over the long run, all workers in the business need to feel part of that business, and take some ownership in the long-term success of the business, its goals and objectives. Simple compliance and behavior manipulation schemes will not succeed over the long run.[19]

But a critical question for all safety practitioners to ask is, "isn't all health and safety really about simple compliance?" For those who have advocated more 'enlightened' approaches to safety management, consider that whether it's about government rules, setting performance standards and accountability for management, or behavior modification, the crux of these strategies, subtly or overtly, is compliance with some type of demand by one individual or group of individuals. Whether it's a soft or hard approach to compliance, the name of the game, to lesser or greater degrees, is compliance.

While safety people have complained about the demands of the regulatory compliance approach for years, what they really want is compliance from their management team and their workers with whatever safety strategy is being promoted or marketed internally. It may be that simple. Then again, it may not be.

COMPLIANCE AND CONTROL: PRACTICAL INCENTIVE STRATEGIES?

In *Brave New Workplace*, Robert Howard states, "When managerial control becomes personalized, the relationship of workers to the corporation is understood in exclusively psychological terms. The very idea of power and control becomes purely therapeutic, a matter of feeling rather than action. And the genuine conflicts of working life—and, indeed, of all social life—themselves become personalized, dismissed as matters of mere individual preference or, worse, social deviance, rather than recognized as legitimate subjects of social and political choice."[20]

While rhetoric of empowerment and team approaches to safety are all the rage, the true picture of what is being promoted may not be as simple as the jargon. In *The Witch Doctors: Making Sense of Management Gurus*, John Micklethwait and Adrian Woolridge note, "For all the talk about empowerment, most are scared, anxious creatures."[21] Indeed, Micklethwait and Woolridge go on to say that, "Nowadays, even the most humdrum jobs can be 'incentivized'. Roughly six out of 10 American companies have a system for rewarding their employees for performance. ... Moreover, incentive pay remains a blunt tool, particularly lower in the organization. In many cases, incentive pay tends to increase paranoia rather than soothe it. If a group of people do extremely well and are rewarded 96% of their possible bonus, the likely response of the recipients will be that somebody has stolen 4% of that they deserved." [22]

Dr. Robert Sass of the University of Saskatchewan notes, "The management professions in North America and elsewhere have inherited—whether an individual manager is aware of it or not—the ideology of Fredrick Taylor's scientific management. This ideology is actually the science of the management of control of the work of others. One of its cardinal tenets is that management dictates to the worker the precise nature in which work is to be performed. In North America, this is exactly what safety is all about!"

It's been said that all safety, in the final analysis, depends on motivation, and that workers need to be motivated to work, and motivated to work safely.[23] Dr. Sass notes that his research on safety incentives indicates the prevalence of a blame-the-victim approach, and studies have refuted the assertion that safety is not merely a matter of motivation.[24]

AN EXAMPLE OF SAFETY INCENTIVES GONE WRONG

The headline of an article on safety incentives in the May 1993 issue of the National Safety Council publication *Safety and Health* reads, "Safety for Hire—Can Companies Pay It Safe?" Comments in the article from several safety professionals highlight the differences of opinion and philosophy behind the safety incentive debate. While the Fox, Hopkins, Anger study may have documented the value of a token economy, in 1993, the chief executive officer of Orrville, Ohio's Wilbert Company commented that after two unsuccessful efforts with the incentive approach, it was dropped as a lost cause. Noted CEO Featherstone, "If employees went a month without an accident, they'd get tickets to a lottery drawing for things like TVs or cash equivalent. ... It went well for a while, but eventually the accident rate would go up again unless we boosted the size of the prize. It just kept growing. Eventually we stopped it after a few years."[25]

This example is just one of many which illustrate the issues faced by those attempting to structure both a short-term and a long-term approach to safety incentives and recognition. In 1971, Wilbert undertook a second attempt to initiate and implement an incentive approach to safety. According to Featherstone, "Exactly the same thing happened.... By 1973, we dropped it. I didn't try it a third time, and have never heard of a successful program that could be sustained."[26] It's precisely these types of experiences and comments which make some sceptics of behavioral approaches stand up and say, "I told you so." Indeed, some would suggest it is the height of paternalism to attempt to be constantly trying to come up with new and innovative ways in which to motivate workers.

Perhaps it might prove beneficial to ask the questions, "What types of incentive approaches work on those who create these approaches for others? Would middle managers react to a token economy the same as a plant floor worker? Would doughnuts do it for your CEO?" It is very often the rank and file workers who have been on the receiving end of many of these safety incentive approaches. It is probably not purely a class issue, but it does raise some interesting questions.

Very often, it can come down to a matter of control. As Alfie Kohn has observed, "Clearly, punishments are harsher and more overt; there is no getting around the intent to control in the 'Do this or else here's what will happen to you' approach. But rewards simply control through seduction rather than force. In the final analysis, they are not one bit less controlling since, like punishments, they are typically used to induce or pressure people to do things they would not freely do—or rather, things that the controller believes they would not freely do. That is why one of the most important (and unsettling) things we can recognize is that the real choice for us is not between rewards and punishments, but between either a version of behavior manipulation, on the one hand, and an approach that does not rely on command and control, on the other."[27]

If we accept the logic of Kohn and others who advocate this point of view, what then become the practical applications of incentives, recognition and rewards? Moving beyond the sterile controlled experiments of academics, what becomes the real world need for these behavior modification approaches? Do approaches to safety incentives make more sense in theory than in practice? And if the phrase behavior modification is offensive to some, might it simply be a matter of changing the word modification to one less threatening? There is a growing body of research and real world experience indicating that not only are incentive, motivational and merit approaches ineffective, they may even be harmful. While generally well-intentioned in their efforts to recognize good performance, they can potentially have the opposite effect.

According to Kohn's examination of the research on incentives, "People's interest in doing what they are doing typically declines when they are rewarded for doing it. ... Scores of other studies confirmed this conclusion."[28] Kohn lays out 14 reasons why, in his opinion, incentives fail. With reference to the United States in particular, Kohn notes, "Consider the countries typically cited as competitors of the United States. Japan and Germany ... rarely use incentives and other behaviorist tactics to induce people to do a better job. This not only debunks the idea that it is 'human nature' to be motivated by extrinsic awards, but also calls into question the uselessness of such rewards, given that these nations appear to be doing reasonably well."[29]

To examine these ideas within the context of occupational health and safety, we need to look closely at the world of work in which we all participate from time-to-time, and to various degrees. More likely than not, we have all been on the receiving or giving end of some incentive or award approach which aimed to get us, or others, to do something. While we may not have been consulted on the approach or had any input what-so-ever on the details, others felt that we would be highly motivated to attempt to earn whatever the carrot was which was dangled at the end of the stick. But it has been said this approach to worker motivation may be one of the biggest mistakes management makes on a day-to-day basis.

In the Winter 1994/1995 edition of the *Small Business Forum*, Peter R. Scholtes, principal author of *The Team Handbook* noted, “The greatest management conceit is that we can motivate people. We can’t. Motivation is there, inside people. Our people were motivated when we hired them and everyday when they come to work, they arrive with the intention of doing a good job. The greatest managerial cynicism is that workers are withholding a certain amount of effort that must be bribed from them by means of various incentives, rewards, contests, or merit pay programs. ...The greatest waste of managerial time is spent trying to manipulate people’s minds and infuse motivation into them.”[30]

CHAPTER 2

SAFETY MANAGEMENT AND INCENTIVE STRATEGIES, PAST AND PRESENT

The history of organized attempts to address workplace health and safety issues clearly shows that identified strategies generally fall into two distinct camps: the focus on the physical aspects of the workplace environment, and the focus on the actions, attitudes and behaviors of those in the workplace and their impact on this thing we call safety.

Through academic studies, research and on-the-job experience, the contrast between the safe place and safe people approaches has been an ongoing debate between safety professionals, governments, unions, and society alike. It's the unsafe act versus the unsafe condition debate. While many feel that either the unsafe act of the workers or the unsafe conditions created by the employers is the prime consideration for safety, the truth of the matter is that it is a combination of these two approaches, together with other influences and factors, which eventually determines the safety of any given workplace, on any given day.

While the focus of regulations and legislation centers mainly on physical conditions and hardware, many seasoned safety professionals say that is not enough, and that it is only if you are able to capture the mind and spirit of those working in any business that you can ever hope to sustain any long-term improvement in safety.

Let's briefly examine the recent history of occupational safety in order to gain some perspective into how the motivational power of incentive, recognition and award strategies has come about.

With more and more workers being injured or killed at work for what were felt to be reasons within the control of management, organized labor began to lobby and push for better and safer working conditions for working men and women. With pressure from labor and the public came the eventual passing of the Occupational Health and Safety Act in 1970 in the US. In Canada, the passage of various pieces of provincial legislation followed soon after.

Generally, labor has fought to have safety addressed through three fundamental strategies: supporting government regulations on safety practices and working conditions, attempting to negotiate safety provisions into collective agreements, and encouragement of the formation of joint labor-management occupational health and safety committees.[1]

Many of the long-held beliefs about the potential for safety to actually impede the progress of business have to do with the perception that safety is not in keeping with the emphasis on production, and that you invariably have to trade off one for the other. The challenge is developing a system in which you get production with safety—not in spite of it, but because of it. This is where the debate about the cost of accidents and injuries has helped the regulators and safety professionals sell the economic benefits of an effective safety management system to their respective constituencies.

It's important to get an appreciation for the background of industrial safety, especially when looking at opportunities for offering incentives for improvement, or considering recognition or reward opportunities for outstanding or exceptional safety performance. Several individuals and textbooks have attempted to examine the progression of the industrial safety movement throughout the years, with a specific focus on the tools, tactics and techniques used by those with responsibilities for accident prevention.

Dan Petersen describes a number of different eras of safety management. When faced with the situation of being responsible for payment for injuries, prudent employers decided it might prove beneficial to address some of the reasons why accidents were happening. While there were differing opinions on the causes of accidents and injuries, as there still are today, at least the recognition of prevention as an aspect began. The move towards inspection of the physical conditions of the workplace led to what's been referred to as the inspection area.

Following the inspection era came the "unsafe act and unsafe condition" era, popularized by the late W.H. Heinrich. It was Heinrich who theorized that some 88% of accidents were the result of unsafe acts. With the move by regulators to cover more work-related illnesses brought about by exposures to various environmental contaminants, the industrial hygiene era was born. Safety professionals of the 1930s and 1940s were looking at a three pronged approach to accident and injury prevention: the physical conditions of the workplace, worker behavior and environmental factors. What Petersen refers to as the "noise era" started after 1951, when a Wisconsin worker filed a claim for hearing loss which he felt was related to his exposure to noise while working in a drop forge.

Following the noise era came the "safety management era" of the 1950s and 1960s, during which safety practitioners seriously questioned the theory and practice of traditional accident prevention techniques, and embarked on bold and ground-breaking research and practice in human factors engineering, management theory and its application to safety, and aspects of quality control, or what we now refer to as total quality management.

The biggest and most influential era came with the passage of the Occupational Safety and Health Act in 1970, and according to Petersen and others, the world of safety management would never be the same again. While the OSHA era may have caused business to refocus its efforts to simple analysis of the physical conditions in the workplace, safety professionals had begun to offer their own perspectives on safety through their efforts, research and grass roots, in-the-trenches experience. In what Petersen describes as the "accountability era", followed by the "human era", the maturity of the safety profession increased to the point where many of the ideas, principles, techniques and program approaches used in safety management programming today came about.[2]

In *Managing for Performance Perfection*, William C. Pope examines the evolution of industrial approaches to safety and safety management, noting the growth of workers' compensation coverage in Great Britain, Germany and the US, coupled with the increased emphasis on regulatory compliance and enforcement of health and safety laws. Frustrated with these simple regulatory approaches to safety and accident prevention, Pope explores the complete history of the safety professional in a modern business, and takes an in-depth look at management theory and practice. In a convincing attempt to show how safety performance can only improve when the safety function is managed as a science, Pope wrote that the understanding of management systems holds the keys in the quest for accident and injury reduction.[3]

With political and social pressure, especially from organized labor, came the many and varied rules, standards and regulations which employers throughout the world have to address in order to be in compliance with occupational health and safety legislation. By and large, occupational health and safety regulations are aimed at providing what many safety professionals would refer to as a minimum standard for health and safety. Even then, it's clear these standards are not consistent throughout the world. Compare the various European common market countries, Canada, the United States and Australia, and you'll see variable sets of standards.

Examine the safety standards of Third World and developing nations and you'll get another perspective on safety. Depending on the economic development and wealth or perceived wealth of various countries, the level of safety differs. Countries, like businesses, can only provide the level of safety that they can reasonably afford. This is not to suggest for a second that safety is not important, particularly when we are talking about life, limb and property, and the value of that life, limb and property. What we are talking about is the level of safety and risk management, which is practical, under the circumstances.

Just what is reasonable, what those circumstances are and how we can create incentives to help promote the desired level of safety is an issue requiring greater examination. As the world shrinks and we all become part of the global village, countries which to date haven't been as active in protecting the rights of working people are being lobbied by the Western world to increase safety standards for many reasons: principal among them, human rights.

Consider the argument put forward by W. Kip Viscusi in his book *Fatal Tradeoffs: Public and Private Responsibilities for Risk*. According to Viscusi, Imposing the risk standards of an economically more advanced society on less developed countries will reduce their welfare by retarding economic growth and the benefits it provides. The main reason for the United States' higher safety standards is not superior awareness of safety's importance but rather a greater ability to afford the luxury of greater safety.[4]

How governments react to these postulates, and subsequently plan their enforcement and safety management strategies will determine, to a large degree, how effective they may be in creating incentives for employers to comply with OH&S requirements. Aside from the moral and ethical arguments that attempt to convince individuals, corporations and unions that safety is important, regulators use incentives to entice, convince or otherwise force employers to comply with health and safety legislation. The traditional (some would suggest least effective) incentive to comply with health and safety legislation is the threat of monetary penalties for non-compliance.

INCENTIVES FOR COMPLIANCE

The scenario goes something like this: government develops and implements legislation, rules and standards. When written into law, they become the benchmarks by which government regulators and compliance officials attempt to determine whether or not employers are concerned with the health and safety of their workers. By and large, the size of the North American workplace means that billions of dollars would have to be spent on compliance officers alone to ensure employers are complying with all the regulations as required. The other practical reality, even for internal safety people, is the degree to which popular safety management strategies like safety audits and behavior-based safety strategies can actually produce the required level of safety.

Safety audits are simply a snapshot in time, and while more effective than traditional measures of injury rates, they also have short comings. While a business may be in total compliance technically, there are still many safety management variables which an audit may not be able to adequately address, such as the safety strategy of the business, current business, socio-economic and regulatory climate, its values and priorities, including culture, organizational behavior (not only at-risk behaviors of workers but those of management and others), and attitude issues (or the soft, safety management issues).

The reality is, health and safety regulations are not absolute because they are not enforced absolutely. However, prudent employers will do their best to comply with legislation due to the fact that the current safety rhetoric says it's good business, and because in the event of a major accident resulting in a serious injury or a fatality, regulators invariably will want to exercise the fullest extent of the law in dealing with an employer who, through errors, omissions, negligence or lack of action, have contributed to the serious injury or death of a worker.

But is it the positive or negative consequences that hold the most potential to get employers to comply with regulations? Similar to the debate about the use of positive recognition or incentives to get employees to comply with safety standards, as opposed to simple discipline for non-compliance, the rationale of how governments create incentives for employers to comply with legislation also needs to be explored.

According to Viscusi's research, while fines imposed on American employers have been negligible (approximately less than $10 million per year), the safety incentives imposed by market forces and workers' compensation were considerably greater.

Viscusi notes, "The wage compensation US workers receive for risks now costs firms over $100 billion per year—a price tag that provides powerful incentives for firms to improve their safety performance and reduce their wage costs. Similarly, the mandated social insurance compensation for workers' under the US worker's compensation system imposes insurance premium costs on firms in excess of $30 billion per year. Especially for large firms, these premiums are linked to the firm's safety performance, providing incentives for safety improvement. Indeed, estimates of the effect of workers' compensation on death risks indicate that workplace fatality risks would be 20 or 30 percent higher in the absence of the workers' compensation incentives."

"Financial incentives do matter, and even from a regulatory standpoint they are a driving force in their promotion of safety. It is for that reason that some economists have suggested that regulatory agencies adopt an injury tax approach to promoting safety rather than relying on the command-and-control regulatory mechanisms. Overall, workers' compensation has roughly 10 times the effect on worker safety as do OSHA regulations. Ultimately, it is the financial incentives created for safety that will affect the trade off which a firm will make in the promotion of safety. Whether regulatory standards or financial penalty schemes are more effective in creating these incentives depends in large part on the magnitude of the incentives being generated." [5]

YOU GET WHAT YOU PAY FOR

The relationship between financial incentives for regulatory compliance and incentives for employers and workers for health and safety should not be overlooked. If employers only get punished for not complying with regulations, and this, for the most part, only occurs after a very serious injury or fatality, how many opportunities does government miss to positively reinforce compliance with safety legislation, rather than simply punishing for non-compliance? The use of positive incentives, financial and otherwise, is a course of action which should not be overlooked by business or government, especially when looking at creative and strategic ways in which safety performance can be improved.

Regardless of the numbers that get bandied about each year, serious accidents and injuries are not common occurrences for most workplaces. The question then becomes, "what is driving the health and safety agenda?"

Over the past 25 years, a lot has been written about the need for new, innovative approaches to address the realities of workplace health and safety. Key to this debate has been the role which staff safety professionals have played in the fostering, promoting and developing a positive and progressive safety culture within their respective businesses.

For the most part, businesses with the resources available to hire full-time safety staff turn to the staff safety professional if there is even the slightest interest in developing a safety incentive, recognition or award program. It is therefore incumbent on the staff safety professional, in an advisory and consultative role, to ensure his or her respective business receives the most up-to-date information, evidence, experience and research associated with both the theoretical and practical application of any corporate approach to safety incentives, or incentive and recognition approaches in general. The key is having competent staff, able to both foster and demand a level of respect for the professional safety function.

Why is this important? To update the old saying, you can't teach an old dog new tricks, unless you know more than the dog! In many businesses, everybody in the company has an opinion of what the safety professional's job should be. Many newly minted safety professionals, just out of school or plucked from the rank and file as a potential candidate for the 'safety guy' because he or she is 'good with people,' suddenly find themselves questioning exactly what their role is in the corporate structure. There may be some periods of uncertainty, but with logic, reasoning, a good job description and the ability to challenge some conventional opinions on the job of the 'safety guy', many safety professionals eventually find their comfort levels and operate rather effectively.

Consider for a second that you are one of those semi-seasoned safety professionals. After a period of less than sparkling safety performance, someone in your company reads a short article in a popular business magazine about the use of incentives to motivate employees to work safely. A light suddenly goes off in his or her head, and the next morning you are sent a short, highly enthusiastic memo, stating that now that senior management has seen the light on safety incentives, XYZ Company needs a safety incentive program too.

You've been asked to put together a program to introduce to all employees next month. The details are not that important. Just get something out there which will motivate the troops, because our safety rates are slipping.

One of the first questions that safety professional should attempt to answer for him or herself is: what is the practical objective of the eventual implementation of this approach? Fundamental to this question is an honest assessment or appraisal of the existing business, regulatory or labor relations climate in that business.

Many books and articles have been written about the importance of culture and its impact on safety performance. Still more has been written about attempting to assess the values, beliefs and attitudes that the business may have, prior to the development or implementation of a safety incentive approach. The reason for this strategy is simple: if you are attempting to introduce an approach to safety incentives or recognition and you're not even sure if the customer for this service, the employee, will buy it, you may be going off on a wild goose chase.

Many may question this marketing analogy, thinking "What exactly it is that needs to be marketed? What's the big deal about getting together a small budget, selecting a few award criteria and passing out the gifts with a smile and a handshake?" The simple answer to these questions may be "nothing." But what are you really attempting to achieve? If, as many have advocated in theory, your intent is to ensure a direct cause and effect relationship between your incentive award and the desired response, action or behavior of an individual or group of individuals, think again.

If your incentive or recognition strategy is to achieve the desired impact, you need to be very clear in the reasons why the approach is being introduced, and what are the desired consequences or expectations. Many safety professionals are asked to develop a corporate approach to safety incentives or recognition, and to have the approach implemented within a week or two. After all, what's so difficult about waiting for a month or two to pass, and if there are no lost time injuries, holding a drawing for a couple of sweatshirts?

While some may consider God to be in the details, others consider strategy, research and analysis to be a waste of time, especially when efforts are directed towards attempting to understand the theory of a proposed approach prior to its implementation. Consultants are often asked to examine, evaluate, audit and analyze the safety management systems of many clients. Invariably, they all want an answer, some quicker than others. Seldom do they want any of the theory, principles or research behind the answers. This is not to say that the homework has not been done, but management tends to prefer consultants be short on theory and long on execution.

But unfortunately, if the approach or recommendation being presented is not based on sound theory, then in reality, the practice will be flawed. So when we examine safety incentives and recognition, we must consider the context within which they're to be used, the existing regulatory or business climate, and how the resources of the organization can best be utilized to maximize any incentive or recognition program.

There are fundamental questions to be asked about the role of an effective safety management strategy in a business that has safety as a central corporate value. Where did that value come from, who has had input into its definition, and what exactly does that value mean to each and every person in the business? In Douglas McGregor's classic, *The Human Side of the Enterprise*, the ability to predict human behavior is identified as an assumption which, while perhaps valid, is nonetheless dependent on the correctness of the underlying theoretical assumption about human behavior. McGregor notes, "We can improve our ability to control only if we recognize that control consists in selective adaptation to human nature rather than in attempting to make human nature conform to our wishes...we will be unlikely to improve our managerial competence by blaming people for failing to behave according to predictions." [6]

Indeed, one of the key, strategic questions which must be asked prior to embarking on an incentive program is, "Are we truly interested in recognizing and rewarding outstanding safety performance, or do we only want other people to do as we say, and do so in the child-like hope that they will be fortunate recipients of whatever gifts, praise or pseudo-glory we decide to dole out?" While the words control or behavior modification may not be palatable to some, especially those marketing approaches to behavior-based safety, the reality is that many safety incentive, recognition and awards programs are simply nothing more than variations on the theme of behavior control.

McGregor recognized this, noting the controlling nature of financial incentive schemes. "Individual incentive plans provide a good example of an attempt to control behavior. ... The practical logic of incentives is that people want money, and that they will work harder to get more of it. ... Incentive plans do not, however, take account of several other well-demonstrated characteristics of behavior in the business setting: (1) that most people also want the approval of their fellow workers and that, if necessary, they will forego increased pay to obtain this approval; (2) that no managerial assurances can persuade workers that incentive rates will remain inviolate regardless of how much they produce; (3) that the ingenuity of the average worker is sufficient to outwit any system of controls devised by management." [7]

STRATEGIC PLANNING: A SAFETY MANAGEMENT IMPERATIVE

Rewarding Achievement-Based Safety

Ask almost any business professional about their business plans and they'll tell you about their marketing strategy, short and long-term investments, product development, including R&D, budget considerations and human and financial resource issues. But it's surprising how little planning is directed towards health and safety performance. I'm not talking about the reflex reactions which invariably take place after an accident occurs, but an annual (or one to three year) plan for strategically improving all aspects of OH&S performance, and setting priorities for what will get done, when it will get done and who will do it.

Most organizations fall into two very distinct categories when attempting to address their safety management issues. On the one hand, there are those who know exactly what they want to achieve because they have undertaken the necessary pre-planning and issue identification long before they get to the point where they are prepared to roll out the activities associated with achieving their objectives. In the other group are those who are not exactly sure what they want or why they want it, but they'll know it when they see it. These are the folks who have undertaken no strategic planning, but hope that with a small program here or there they'll achieve their objectives (if they have even gotten as far as defining objectives).

Let's look at some very practical, real world examples of strategic plans, with particular emphasis on opportunities for creating safety incentives or subsequent recognition opportunities. Then let's look at applying the same logic to your safety management activities.

A consultant has completed a contract for a client who wished to have senior executive safety education performed. It was felt that senior executives had not been active in health and safety management issues, and staff felt that senior executives needed more awareness of OH&S issues and due diligence. Part of this process included trying to get a handle on what objectives the client wanted to achieve with their safety management strategy. In other words, "What do you want to do and how are you going to go about achieving it?"

After some discussion and the sharing back and forth of questions on training objectives, the client and consultant were very clear on what the objectives were to be. The consultant was to conduct a standard series of sessions on safety management and due diligence issues for senior management, and subsequently facilitate a safety action plan session which, when completed, would enable the client to address specific, strategic safety initiatives to help move the organization forward from a health, safety and environmental perspective, up to and including the year 2002.

The objectives were very clear: deliver a consistent safety message, ensure people understood their role in an effective safety management system, and propose short and long-term actions for future improvements of select safety initiatives. Part of this process was to include an executive safety recognition and performance evaluation option to recognize those who met specific prevention initiative targets.

Another client, with a number of unionized employees at one particular facility, felt there was a problem with safety that couldn't continue. Aside from poor statistical performance, it was felt there were a considerable number of problems with their existing safety system and culture. These opinions were mostly subjective, but as the saying goes, "where there's smoke, there's fire." It was eventually determined that there was indeed some reason for concern.

The consultant was asked to attend a meeting with senior management and union representatives to try and figure out what was wrong with the safety system and how it could be fixed. Without having much information about their safety system prior to the meeting, the consultant was challenged to try to identify, during that meeting, what some of the issues or problems were.

What the consultant decided to do was structure a series of questions that would be discussed. These questions were based in part on prior safety perception surveys, and were intended to obtain sample feedback on the culture and safety maturity level of this facility. The manager of this facility also pulled a safety incentive award idea from his hat at the meeting. Something about giving out a fire extinguisher every month during which the employees didn't have a disabling injury. He just didn't get it.

So what was the difference in the strategic planning for safety by these two clients? Simply put, Client A knew what they wanted, why they wanted it, and how they were going to achieve it. The consultant was simply going to facilitate their awareness sessions to help improve the knowledge level of their executives on health and safety management. Client B felt they had a problem, but they didn't know what it was or what they were going to do about it. They weren't even too sure about why they had contracted the consultant to attend the meeting, other than hoping that he would have some magic answer to their problem.

Let's review the two scenarios. If you were the consultant with Client B, what would you recommend? If your recommendation would be that they first attempt to identify critical problem or opportunity areas, you'd be following the same strategy the consultant proposed. That process can be as broad as it is long, using audits, perception surveys, interviews with managers and workers, physical conditions assessment, and other system appraisal tools. You can't fix a problem or improve a situation until you first define what the problem or opportunity is. And you need to make that determination based on fact, not speculation, opinion or value judgment.

Once you identify the key issues, it then becomes a matter of identifying the priorities, or the critical issues. Depending on your priorities (or those of a client), those critical issues can be compliance issues, management or cultural issues, or risk management issues. The decision should also depend on where the most potential for improvement exists. With this information in hand, it will be much easier to set specific prevention targets or initiatives, and provide incentive or recognition options or opportunities for those that make the mark. While this is a strategic planning process which gets input and buy in from all operational levels, and assigns suitable recognition appropriately, remember that there is no simple means of making people comply or forcing individuals to comply with your opinion of correct behavior.

Whether you're a staff safety professional or a contract consultant, the methodology for attempting to identify strategic safety improvement opportunities is the same. First of all, what is your role in the safety improvement process? Are you simply going to be asked to give your opinion, or are you expected to help identify the issues? Are you going to be involved in any training or professional development (on the receiving end) or are you being asked to facilitate the improvement process?

There are many changes currently taking place in most workplaces throughout the world. These are not safety changes, per se. They are changes in our economic, social and political structures, and they are having tremendous impact on current organizational climates. Some of these changes have implications for safety incentives and recognition alternatives as well as health and environmental issues.

Those who simply wait until these changes filter their way down to the issues that can impact safety performance may find that they do not like the changes. But it is possible to be prepared. The issue is not resistance to change, but change management. From a safety management perspective, those who do the best job in strategic planning with what they know can be better prepared to deal with what they don't know.

Risk and Reward: A Risk Management Approach

Strategic planning efforts need not take days or weeks, nor need they be turned into a bureaucratic process. Committees are fine, and there is a time and place for committees. But an effective organization cannot be managed by committee. (Who was it that said a camel is a horse designed by a committee?)

Even team approaches to strategic planning may not be suitable for a team-based effort. Consultation, yes. But very often, because of the desire to have every activity perceived of as an extension of a team-based philosophy, we bog teams down with a lot of unnecessary effort.

The current fascination with teams and teamwork doesn't necessarily mean that everything from a safety management perspective needs a teamwork approach. Individual performance matters a great deal. Again, here is where a little strategic planning can pay off big time. The larger the organization, and the risks associated with that business, the greater the potential for success from a detailed strategic planning process.

To help determine the focus of the strategic safety planning process, consider the following:

1. What are the risks associated with your business/industry?

2. What has past performance indicated (i.e., high frequency, high severity, total accident costs, workers' compensation costs, audit findings, other performance measures or indicators)?

3. How would you rate your safety maturity level? In other words, are you simply doing the bare minimum, or have you identified safety as a strategic, core business value, and integrated or made attempts to integrate safety performance into business improvement efforts?

4. How knowledgeable is your executive group about current safety issues, practices and trends? Are they familiar with due diligence concepts? Do they play an active role in safety initiatives, or are they simply on the sidelines as cheerleaders?

5. What do you want to achieve over the short and long term? Where do you want to be this time next year? How about in the next one to three years? Where do you see your organization in the next five years?

While some of these targets may seem a long way off, it's important to remember that the process of planning how you intend to get there is just as important as setting the actual target.

It's also important to remember that the short-term goals need to be balanced with long-term objectives. It's simply a matter of asking, "Is safety important in this company?" If the answer is yes, then the next three questions should be, "What exactly is it about safety that's important, what do we have to do to achieve our objectives and who is going to do it?" Even more important is the execution of your strategy, and integral to that strategy is whether you plan to offer incentives to prompt action and reward and recognize those who have played an important or pivotal role in helping the strategy become reality.

Even more to the point, who in your business can actually play a meaningful role in helping improve safety performance? Is it simply the worker through safe behaviors? Is it supervisory and other line management through their leadership in day-to-day safety initiatives? Is it senior management and their strategic planning focus that gives the organization its safety improvement goals and targets? Or is it all of these and more? In order to achieve an effective incentive and recognition process, you need to develop a synergistic safety incentives model, a model which involves all levels of your business from top to bottom.

It's been suggested that reinforcers are actually more effective than rewards, due to the fact that most rewards are usually in the future, and there may be a fair degree of uncertainty about how they will be give given.[8] To determine which reward or recognition option best matches your values and beliefs, and the value and belief system of your business, it's important to realize that there are various practical options available. While generally thought of as common sense, recognition is something that must be actively thought about due to the fact that for most of us, it is not normal behavior.[9]

Just as there are many and varied approaches to safety management, there are a wide variety of options for safety recognition. These can be classified as left brain or thinking approaches to recognition (Figure 2.1.) and right brain, or emotional options. I'm a thinking recognition advocate. But the key is to determine which option is right for your business. What about you? Which side of the brain do you prefer?

Now have a look at what have been deemed the softer qualities for recognition (Figure 2.2.) and see if you can objectively place yourself in one group or the other, or whether you feel the need to pick and choose from each.

FIGURE 2.1.
Thinking (Less Personal) Recognition Options

Thinking
Using Your Head
Keeping Your Cool
Offering Constructive Criticism
Achieving Precision
Being Fair-minded
Standing Firm
Being Objective
Upholding Justice
Not Taking Things Personally
Being Analytical
Maintaining Principles
Truthfulness
Problem Solving

FIGURE 2.2.
Feeling/Emotional Recognition Options

Feeling
Subjectivity
Caring
Appreciation
Persuasiveness
Warmth
Gentleness
Using Your Heart
Creating Harmony
Empathy
Taking Things Personally
Upholding Values
Tactfulness
Cheerleading

If you can answer these questions honestly and openly, and can bring key players into the strategic planning process, your short and long-term safety improvement plans stand a better chance of success. Here too is where the strategic planning part of your existing safety systems and 'culture' will pay off.

In terms of accurately reading the mood of the people on your business, are the key players more thinking or more feeling or emotional? How can you find out? Just as you would conduct a perception survey to gauge the general feelings and thoughts of your workforce to determine their likelihood of accepting a new or innovative approach to accident prevention, so too can you conduct perception surveys to determine which approach to safety incentive, recognition or reward schemes may stand the best chance of success.

To conduct the safety incentive perception survey, begin by asking some very simple, honest questions which will be used to help guide the efforts and direction of your strategy. This is also a very practical way in which to generate both interest and buy in to your approach, as the logic and rational for your approach can be, and should be, based in part of the focus and direction which your perception survey identifies. After all, the people whom you hope will accept the approach are those whom you are surveying.

If you can come up with some answers as to what the people in your business are more likely to accept, you'll stand a better chance of selling that approach, as opposed to pushing it down—it's an oxymoron.

Get an honest and sincere sense of what makes your people tick, and you stand a better chance of developing an incentive or recognition system based on their needs, desires and values, and you'll also gather considerable amount of detail about other key aspects of your safety culture in the process.

CHAPTER 3

MOTIVATION AND LEADERSHIP

When you examine the safety management strategies of other businesses, how often do you see strategies which are developed as a result of shared beliefs, values and principles? More to the point, how often to you see true commitment of all workers, from the CEO to the shop floor, in helping set and define the safety strategy for the business? While there has been much talk among business, labor and government about the need for ownership of health and safety issues at all levels of business, getting beyond lip service, rhetoric and cliché poses a challenge for many.

In most conventional safety systems, the traditional forms of command and control are still very evident, in spite of all the talk about empowerment, and especially as reflected in the safety reward schemes presently in place. To the cynical, and perhaps even those who have seen many of these new approaches to management theory come and go, empowerment is simply another way in which others may be able to do much of the work and be held responsible if something goes wrong.

To folks who may lack street smarts, it may be seen as a way in which workers can learn how business decisions get made. To the practical minded, it has been a struggle to deal with management not wanting to give up their traditional power base: if management no longer has the ability to use the tools of command, control and coordination, why is management required at all? The reality is that this question is not only being asked more and more, but more and more managerial workers are finding out that they are not needed.

Today's workplace is changing rapidly, and will continue to change at record speed. Many senior executives, frustrated with the lack of flexibility and inability of many workers to march to the beat of the new CEO's drum, complain loudly that workers don't like change. They complain that workers have gotten too comfortable and aren't tough enough to meet the rigors of competing in the new economy. This is highly paternalistic and arrogant.

Workers don't mind change, they just don't like being told to change. Worse still, they feel helpless and ignored when they haven't even been consulted on change, and how that change may impact them, their jobs and their families. This is borne out by a recent survey by the Canada Health Monitor that found that 25% of workers reported mental or emotional health problems associated with their work, as opposed to only 9% who indicated they had experienced a workplace injury. The study noted, "It is not necessarily change people have difficulty with, it's the uncertainty and loss associated with the change. ... Today's unstable work environments are demanding from workers a flexibility many have not developed, coupled with increasing job expectations." [1]

In order to be able to deal effectively with the issues stemming from the fast-changing world of work, employers need to give serious consideration to the corporate culture that has contributed to or created the levels of stress and anxiety for people at work. Fundamental to addressing these problems is a critical examination of the corporate strategy for survival and growth, as well as the corporate strategies for occupational health and safety which, hopefully, should be integrated into the overall business plan.

It's interesting to note that central to the focus of almost all traditional approaches to safety incentives is the reduction or elimination of occupational injuries in order to be eligible to receive a prize. However, as the Canada Health Monitor study points out, breaking rules and getting injured are not what is causing many of the occupational health and illness problems in the workplace. This is one more point that weakens the arguments of injury-free period incentive advocates, and especially those who favor simple rules or unsafe worker behavior incentive approach.

With traditional command and control management styles, many employees don't feel part of the process of helping the business stay in business. They have no intellectual ownership of the business, its goals or objectives. So too are many of our approaches to health and safety management. While there are lofty executive goals set for corporate health and safety performance, or safety staff develop theoretical safety strategies based on the latest safety fad articulated by some academic guru, many workers are feeling left out of the safety equation. To them, safety is merely the fact that many of their behaviors are being documented and targeted for feedback on a daily basis, in return for pseudo rewards and praise.

A SHARED APPROACH WITH COMMON GOALS

If you want someone to buy what you're selling, you're got to be able to do one of two things: convince them that they need what you're selling (whether they do or not may be irrelevant), or help them define their needs, and then sell them what they then know they need.

Let's look at the issue of ownership in health and safety management more closely. There are better ways to create shared opportunities and common goals for all employees involved in the safety process. I'm not simply talking about your average shop floor worker. I'm talking about a shared, strategic approach to health and safety management which, by design, may start at the top, but which cascades down throughout the entire business, with positive benefits for management and worker alike. This should be deemed a shared approach to health and safety management, with shared responsibilities and accountabilities. There should not be an emphasis on one group to the total exclusion of all others.

THE INTERNAL RESPONSIBILITY SYSTEM FOR OCCUPATIONAL HEALTH AND SAFETY

As this book was being written, there was an interesting debate taking place in Canada, particularly with respect to the current state of affairs of occupational health and safety. The issue that was driving the debate was the focus of health and safety legislation in Canada, juxtaposed with the views of many on the popularity of the behavior-based approach to safety. Central to the Canadian discussion were the findings of the inquiry into the Westray coal mine explosion discussed earlier in this book. The fact that the Westray coal mine disaster was a great tragedy cannot be denied. But for those who delved into just how and why the explosion occurred, there were many interesting and important lessons to be learned, especially from the point of view of responsibility for safety, and the role which incentives or bonuses may or may not have played in the disaster.

The Canadian system of internal responsibility for occupational health and safety has been patterned, to a degree, on similar systems used in the UK and various Scandinavian countries. In most countries, government regulators hold themselves up as being the protectors and benevolent benefactors of workers from a health and safety perspective. However, it's interesting to note that while many in government are very quick to take the credit when industrial safety performance improves, they are not quite as anxious to step forward and take responsibility for industrial safety performance when it is poor.

The US may be able to take some lessons from European and Canadian counterparts in terms of occupational health and safety. In Scandinavian countries, the seemingly oxymoronic term self-regulation has been guiding the development and implementation of workplace health and safety strategies for years. While many businesses struggle to come up with new and wonderful incentive approaches to get workers to perform some activity to which they may be naturally averse, the system of internal responsibility holds much promise for worker motivation, empowerment and recognition from a health and safety perspective.

Generally speaking, internal responsibility requires both a shared and specific responsibility for occupational health and safety in the workplace. While many safety programs are littered with all manner of slogans and clichés, perhaps the one which best typifies the supposed approach to responsibility is "safety is everybody's responsibility." To a degree, this may be true, but if everyone is responsible for safety in general, nobody is responsible for safety in particular.

In the internal control system, the name of the game is ongoing system documentation, reviews and regular input from employees. The whole thrust behind the idea is that workers and management participate in the development of the system which serves their own best interests. The core of the internal responsibility system is that you have to continuously work at creating a safer environment. With a heavy emphasis on the principle of employee involvement and ownership as part of the internal control system, Norwegian companies implementing this approach have noted increased safety and health awareness, in addition to improved risk assessment.[2]

One of the potential benefits of the Canadian approach to the internal responsibility system is that in some jurisdictions, it spells out actual responsibilities for occupational health and safety, from the CEO and other management and supervisory personnel, to workers. The value of this clear definition of safety responsibilities is that it can accommodate both high level and lower level safety incentive or recognition approaches. While many approaches to traditional safety incentives have the worker as the only focus, the utilization of an effective safety incentive and reward system can ensure that, strategically speaking, all human resources within the organization play a part and make a contribution to improved safety performance.

To help put this in perspective, let's look at some of the traditional approaches to accident prevention and theories of accident causation and examine effective approaches to safety management systems and complimentary reward strategies.

DESTROY THE SAFETY MYTHS AND CLICHES

Most of us have strong thoughts and feelings about occupational health and safety. Indeed, if we didn't, we probably wouldn't be in the safety business. Having said this, there are times when our strong feelings get in the way of rational thinking and scientific evidence. These emotional responses typically manifest themselves in clichés and slogans that, over time, become a foundation of myths upon which safety gets built. The problem is that myths and clichés seldom stand up to scrutiny, and eventually, down comes your safety initiative or incentive approach like a house of cards.

As we've seen, safety issues get dealt with on any number of levels. There are legislative compliance and due diligence issues to consider which leave little to the imagination. The objective is to comply. If you have limited time and resources, compliance is nothing more than one group of people making sure another group of people do exactly what they are told. There's not too much room for creativity in this scenario.

But, there are also scientific and engineering principles that address occupational health, accident causation theories and behavioral issues. Add organizational behavior, corporate culture, current business climate and the regulatory perspective in which your business operates and you've got a very interesting mix. These factors force us to be a bit more objective and less reliant on value judgements. They also force us to do some really hard work, to go beyond and mere compliance.

The safety movement itself has its own history, folklore and myth. Over a period of years, some of these myths have developed into a type of conventional wisdom, which might be more accurately referred to as unconventional wisdom. A lot of what we do in safety is based on conventional practice or tradition. Too often, this convention or tradition is based on ill-defined logic, rationale or science. As Richard Zeckhauser of Harvard

University noted in his assessment of W. Kip Viscusi's book *Fatal Tradeoffs*, "The domain of risk analysis (safety) is too often characterized by soaring and obfuscating rhetoric supported only by snake oil analysis." [3]

So what are some of the characteristics of this "soaring and obfuscating rhetoric?" In safety, they are often apparent in the clichés and slogans that are used to rationalize safety efforts. This falls way short of any lasting impact. So what can you do to recognize the "snake oil analysis" from the logic and science? Read through the following list of safety slogans to see if they reflect opinions held by people in your organization, and then let's look at the supposed meaning of these slogans and myths, and their shortcomings:

1. Safety first.
2. Safety is your number 1 job.
3. The main cause of accidents is worker carelessness.
4. The supervisor is the key to safety.
5. Safety is everybody's responsibility.
6. Most accidents are caused by the person who had the accident.

These myths and clichés are based on a variation of those identified by the late Dr. Ted S. Ferry of the Faculty of Safety Sciences, University of Southern California. Ferry noted, "After reading this critique of common safety slogans it may seem that nothing is left on which to build a safety program. If slogans, clichés, or shock devices are what constitute the program, then there is not much to start with and change is needed. If these items were major portions of your effort, then the depth of your program is just that—slogans, clichés, posters and one-liners—myths of safety. Look at it from a business viewpoint. Would slogans, truisms, clichés and the like increase your profit; improve quality control; make shipping do a better job;

encourage staff to more careful planning? The answer is no, and neither can it be a basis of a successful safety program." [4]

Yet many traditional safety incentives are still used as the tool of choice to force, cajole, motivate or otherwise convince workers that if you don't break any rules—or more to the point, if you don't sustain an injury—there will be a prize waiting for you, or your group, at the end of the month.

If you continue to base your safety performance efforts on out-dated safety clichés and slogans, you can never hope to have efforts aimed at improving safety taken seriously. Slogans like, "You are the key to safety" and "Employees should be more careful" only serve to perpetuate the myth that the worker is the one who needs to be encouraged, guided and motivated to look out for his or her own safety. It says nothing about safety training, management systems, information technology, risk management, or the other variables a modern safety management system has to consider to assist in identifying and appraising loss potential or strategic planning for safety. These factors need to be considered in any reasoned, intelligent approach to safety.

My tolerance for slogans and clichés is minimal at best. At worst, they're highly inflammatory and paternalistic, and can lead to a total contamination of the safety improvement effort. Let's look at the slogan, "Safety is your number one job." For safety professionals and consultants dependent on safety services to put food on the table, safety is our number one job. But when you look at your own business or institution, ask yourself one simple question: What is the main reason your organization is in business? Answer that question, and you have a better understanding of what your number one job is.

If, for example, you work for a tire manufacturer, your number 1 job is to help produce a quality product (tires) for your customers, at the best possible cost, safely. Safety is not your number one job, but it is no less important than any other job that you have. Sort of a first among equals. If you work for a regulated electric utility, your number one job, and the job of your organization, is to produce, distribute and sell power to your customers, providing quality, reliable service at the lowest possible cost, safely. If your company sells breakfast cereal ... well, you get the idea.

When we say safety is first, what happens? What comes second? Production? Quality? Costs? Invariably, we put people in the position of having to make a choice. This is unfair, and unrealistic. If we have to depend on slogans, at least try to say something like, “Safety & Quality: the two go hand in hand.” The goal should be to attempt to integrate safety into all other management actions. No more important, no less important. Believe me, this challenge will be a whole lot greater than coming up with a new safety slogan.

If you still insist on using cute clichés and one-liners in an attempt to catch the imagination of your workers, consider the slogan “you are the key to safety”. Most keys are on a key chain, and that key chain is held by someone who turns the key, and for the key to do its job, the rest of the machine into which the key is inserted must be functioning efficiently.

If we believe what the management gurus tell us, safety performance is not dependent on the programs that an organization conducts, like HazCom or WHMIS, investigations, inspections or safety meetings. The gurus tell us that successful safety performance depends on the safety culture that exists, and the values that an organization has for safety.

If your organization says it values safety, and part of your existing culture is that you use lots and lots of slogans and posters, what does this say about the potential for safety performance in your organization? For years, research and theory suggested that most accidents were caused by unsafe acts of people. This was an axiom of industrial safety put forth in 1931 by W.H. Heinrich, considered by many to be one of the founders of modern industrial safety. This theory has not been challenged to any great extent, and in fact, this theory—true or not—has been treated as gospel and interpreted as common sense.

Consider what some of the safety gurus and theorists are suggesting. Let's look at a variation of the unsafe act scenario. In Petersen's book, *Safety Management: A Human Approach*, he notes, "In most cases, unsafe behavior is normal human behavior; it is the result of normal people reacting to their environment." Management's job is to change the environment that leads to unsafe behavior. An unsafe act, an unsafe condition, and accident: all these are symptoms of something wrong in the management system.[5]

It's been suggested that the original research conducted by Heinrich was fundamentally flawed in that his conclusions about the unsafe acts of workers were based, to a large degree, on the opinions of management representatives who were interviewed and asked their opinion of the causes of accidents and injuries. It is safe to say that management may have been less than inclined to implicate themselves in the accident causation model, and some would suggest that it would be only natural, yet also inaccurate, for management to default to the unsafe worker scenario. Many labor unions have attempted to use this argument to attack behavior-based approaches to industrial safety.

So where does this leave our myths and clichés? If accidents and injuries are symptoms of management system defects, then the roots of prevention must, by definition, be based on a management system or systems, and with management itself. Now rationalize this with management extolling workers to "be more careful and keep safety uppermost at all times" (how often have you seen this written in an accident investigation) and you can see the myths are still alive and well.

If accidents are symptomatic of management system problems, yet management keeps telling workers to be more careful, what sort of dysfunctional approach have we created? Most of us continue to use these tired, shop-worn clichés to help spur safety initiatives. We've seen it time-and-time again from industry, labor and government. We've seen it repeated ad nauseam in safety training videos, especially those with titles like "Safety is Everybody's Responsibility." Again, if safety is everybody's responsibility in general, then it's nobody's responsibility in particular.

If safety is to work, very specific responsibilities need to be defined, with very specific levels of accountability for those responsibilities. This goes far beyond mere behavior observation and modification approaches. This requires a total safety management approach from top to bottom. If you're prepared to take the safety incentive and recognition debate to this level, you'll at least be preparing the foundation of a systematic, total approach to safety awards, with all players sharing in the identification of high level goals and objectives, performance targets, risks and rewards. An achievement-based safety culture is required to replace the simple unsafe behavior schemes and reward plans that characterize many safety systems in operation today.

What about another cliché, "We need more safety awareness?" Some never achieve awareness, yet they still wonder why safety is not working. Let's look at the Safety Action Model that I advocate for my clients, which takes simple, passive awareness to ultimate action:

1. Information.
2. Awareness.
3. Knowledge.
4. Skill.
5. Commitment through policy, responsibility and accountability.
6. Action, including performance measures, targets, milestones and recognition options.

We are inundated with more information than we know what to do with. A simple 10-minute surf on the Internet will verify this. Simply saying that information will increase our awareness may be accurate, but that awareness may never result in increased knowledge and skills. However, if it does, safety knowledge and skills should be channelled through a workable policy, safety responsibilities and clearly defined accountability. A lot of us never get past point number two.

If there is one great failing of industrial safety, it is the failure to ensure accountability. As Petersen has noted, "Instead of simply preaching for 50 years that the line has responsibility, we should have been devising procedures to fix such accountability. When someone is held accountable (is measured) for something, responsibility will be accepted. Without accountability there is no accepted responsibility." [6]

If government really wants to aid industrial safety performance, it should work on this for the next 10 years and see what difference it makes. Industry, labor and governments need to partner to help develop safety management systems that ensure accountability. And not simply punishing business for not complying with standards after a serious accident takes place. This may not sound easy, and indeed it may very well not be easy, but it's time we put our money where our mouths are and help create the systems which will determine successful safety performance. We can then reward for results, not simple luck or forced cooperation. Rewards and recognition based on strategic approaches, with dedicated and committed resources, can put safety miles ahead of the paternalistic, do this and you'll get that approach.[7]

Back to our myths. How many of them does your business subscribe to? How hard have you tried to destroy those myths? Perhaps you've marketed some of them, or perhaps these myths are still being perpetuated by your executive or supervisory group. Here's the challenge: Work with your organization, management and workers, to help destroy these myths. They only serve to perpetuate the notion that the worker is the only one who needs to be encouraged, guided and motivated to look out for his or her own safety. They are based on pseudo-psychological principles aimed at addressing worker behavior exclusively. And we've learned that safety is eventually determined by the behavior of everyone within the organization, not just workers.

Think about it. One safety myth destroyed can open up opportunities to get some real safety objectives accomplished, and help lay the foundation for continued successes based on logic, reason, fact and science, not folklore.[8]

THEORY AND REALITY: WHAT ARE THE CAUSES OF ACCIDENTS AND INJURIES?

As previously noted, there is documented evidence of a dichotomy of opinion when it comes to theories of industrial accident causation. The over-simplification of narrowing accident causation to the unsafe act-unsafe condition variable contributes to one-dimensional thinking on the part of regulators, business, labor unions and safety professionals, especially when it comes to taking sides on the why accidents are still happening debate. If we accept to any degree the notion that accidents and injuries are the result of actions or in-actions of people, we should assume that a multi-dimensional accident causation model may be appropriate, which includes the president and CEO, the shop floor worker, and all other employees and management in between.

Heinrich and his advocates may have been correct in their observations, based on the information and research available to them at the time, but the world has changed considerably since Heinrich's unsafe act theory was new. One of the more reasoned and intelligent examinations of the workplace accident scenario came from a 1974 report of the Ham Commission. In his *Report of the Royal Commission on the Health & Safety of Workers in Mines*, Commissioner James M. Ham noted, "The Commission believes that emphasis on unsafe conditions and unsafe acts falsely dichotomizes and generally oversimplifies the organic circumstances out of which accidents arise...."

The report continued, "In the hearings before the Commission there were two particular points of emphasis in relation to accidents, unsafe conditions and unsafe acts. Some workers' representatives emphasized the former, and some management representatives the latter. Unsafe conditions may have their origin in unclearly defined and communicated management objectives. They may arise through defects of plant and mine design, through methods of work inadequate in themselves or inadequately supervised, and through tools, equipment, and processes inadequately maintained. Unsafe acts of any person may originate in want of vigilance, training, skill, physical strength or judgement when all conditions of work are otherwise within standards...."

Ham concluded, "The apparently common view that the great majority of accidents are the direct result of nothing more than unsafe acts or unsafe conditions is, in the Commission's opinion, too restricted a view of the human problem of accidental injuries. Workmen and their supervisors at every level may act unwisely, but they do so within a system for the performance of work whose responsibility it is to set clear and supervised standards of what is expected." [9]

In *Professional Construction Management*, Donald S. Barrie and Boyd C. Paulson debate the role which occupational health and safety plays in the management of engineering and construction projects. They note there are a number of motivators for improved safety performance in any business or construction project. Principal among them are humanitarian concerns, economic costs and benefits, legal and regulatory issues, liability consequences, and organizational image.

While there are many and varied ways in which to measure the effectiveness of safety performance, many still use disabling injuries as the prime health and safety indicator. According to Barrie and Paulson, behavioral approaches to health and safety need to be explored, in as much as they cited that studies have shown that roughly "80 percent of all industrial accidents result from unsafe acts in the accident chain, and not just from unsafe conditions." [10]

Interestingly enough, those studies are never referenced or documented by Barrie and Paulson, but when they refer to the behavioral approach to safety, it is not simply the behavior of the worker that they reference. The behavior of management and supervisory staff is also deemed to be important. The key point here is that if, as many in the total quality management (TQM) movement have stated, unsafe acts of workers are the main cause of accidents and injuries, it is merely symptomatic of other safety management system problems or dysfunctions which must be addressed and fixed before the behavior of others can ever hope to be positively impacted.

Barrie and Paulson give an example, based on research conducted at Stanford University's Graduate Program in Construction Engineering and Management, of a more wholistic approach to behavioral-based safety which considers the many different aspects of an effective safety management structure, not simply the behaviors of workers. In four separate and distinct studies, they focus on a behavioral approach centered on top management, superintendents and project management, foremen, and workers.

Quoting from a study conducted by Dr. Raymond Levitt, Barrie and Paulson noted the following actions, with particular reference to safety incentives, which senior management can employ with respect to reducing accident costs:

- Know the safety records of all field managers and use this knowledge in evaluating them for promotion or salary increases.

- Use the cost accounting system to encourage safety by allocating safety costs to a company account and allocating accident costs to projects.

Also recommended for top management was the discriminating use of safety awards, suggesting that safety awards for workers, if used, should be incentives (awards of nominal monetary value), based on first-aid injuries rather than on lost time accidents. In addition, it was recommended that safety awards for field managers should be bonuses (awards of substantial monetary value, made in private) based on lost-time accidents or insurance costs.[11]

The key issue in this approach is that simply having worker behavior as the focus of a behavioral safety approach is narrow in focus, superficial in addressing total safety management system issues, and neglects to include or involve a very important element in an effective safety management strategy: management itself.

LEADERSHIP: A SAFETY MANAGEMENT IMPERATIVE

Leadership is one of those buzzwords which comes up from time to time to reflect how we feel about institutions, business or people. How often have you heard people talk about leadership in government, the military, labor unions, business executives, and even supervisors? Volumes have been written about the subject, with opinions and views differing so widely that you'd think this thing called leadership was more of a hope and a prayer than a real life attribute.

Safety needs good leadership. There's generally no disagreement on this point. The rub comes when you ask people how this leadership can best be demonstrated. In the May/June 1993 issue of *Canadian Occupational Safety*, I wrote, "Safety professionals who look inward for solutions to existing and future safety problems are looking in the wrong place. The world does not revolve around safety. In fact, a considerable number of so-called safety issues are simply reactions to other key issues and challenges that our society has been faced with lately. Continuing to react and plan safety improvement initiatives only around the so-called traditional safety issues will result in a phenomena known as inbreeding—limiting your search for new ideas and methods within your own industry."[12]

In attempting to come to grips with an approach to safety incentives and recognition, it's vitally important to ascertain exactly what your approach is attempting to achieve. Traditionally, organized labor has played a strong leadership role in lobbying governments to pass more and tougher laws for the protection of worker's health and safety rights. Indeed, strong regulatory enforcement is the main trump card which, in the opinion of organized labor, needs to be played strategically in addressing occupational health and safety issues. But while the moral higher ground in terms of the ethical approach to safety still has merit, serious students of the health and safety profession need to critically ask whether there is more to this equation than meets the eye.

As noted earlier, while everyone generally agrees that superior safety performance should be recognized, the challenge for many is clearly defining, and then articulating, just what constitutes superior safety performance. If we accept the behaviorist's theories, we will perhaps conclude that it is the unsafe or risk taking actions or behaviors of workers which need to be measured, quantified and qualified with respect to how safe or unsafe they are, as demonstrated by the presence or absence of risk taking behaviors or the traditional unsafe act. Feedback to workers will tell them how well or how poorly they are performing, and provide positive recognition or reinforcement to those which meet the behavioral expectation. If only if was that simple, and would that all workers would simply respond to what some critics refer to as a seemingly paternalistic approach.

While the behavioral approach has merit, can result in select worker behavior improvement, and may be able to achieve some measurable objectives in the short term, a simple worker behavioral approach fails to address the entire safety management system under which many businesses operate. Key to this approach is the behavior of senior and middle management, and first line management as well.

Many who have written on modern approaches to safety management continue to use phrases like a "shared approach," "team-based safety," "common goals and objectives," "employee ownership" and "worker empowerment." If we are to look at any rational approach to safety incentives, we should make sure that the opportunity to participate in the strategy to improve safety performance runs vertically from the top to the bottom of the organization.

To merely focus on worker behavior ignores those who have the most opportunity to control and regulate all the means of production: management. Having said that, I would very much like to see some empirical evidence which would suggest that the same factors which motivate workers to perform safely through the exhibition of correct or safe behaviors, also motivate middle and senior executives to attempt to attain superior safety performance.

Perhaps what we are really talking about in terms of how traditional safety incentive and recognition schemes have worked in the past boils down to a type of class structure, whereby some may feel that the lowly worker needs only simple prizes and token recognition for performing, in a way which much resembles how we train pet cats and dogs to sit up or roll over on command. Indeed, the word grunt is often used to refer to workers, who, for the most part, have been the target of many and varied approaches to incentives, motivation, behavioral modification and other performance improvement strategies.

Pick up any popular management theory textbook and you'll be assaulted by a litany of new and improved approaches to business and the science and theory of management. By way of example, to promote a health and safety conference in New Brunswick, Canada in 1994, conference organizers chose as their theme, "Shifting Gears in Health and Safety" (perhaps made in reference to Nuala Beck's best seller, *Shifting Gears: Thriving in the New Economy*).

The conference brochure noted, "The changing economy and workforce downsizing mean managers expect more, while workers are stressed out from job pressures, increased workloads and lack of control over either the job or the future. Companies are struggling to do more with less, and still be competitive in ever-tighter markets." A tall order indeed faces the health and safety community today.

With all these pressures on health and safety, which factors will make the difference? Time and time again, it has been shown that one of the main factors that help industry, government and labor out of a quagmire is leadership. This leadership role is not the exclusive domain of management. It can come from anywhere. Unions can play a key leadership role, as can government. So too can safety professionals. The key is knowing what to do and how to do it.

I'm not talking about repeating all the old safety clichés that have disguised themselves as leadership qualities over the years. What I'm talking about is absolute, honest-to-goodness leadership. A clear focus on what needs to be done, how best to do it, and the will and desire to get on with the job. Those who can best articulate these issues and then not only talk-the-talk, but walk-the-walk will emerge as the safety leaders in the decades to come.

Leaders need to be acutely aware of which issues have the potential to impact their businesses and the employees who work in them. Leaders also need to be aware of how ineffective health and safety systems and poor safety performance can negatively impact their businesses, and subsequently, their profits and potential rates of return on shareholder investment. Why? Because the issues shaping the present and the future will often dictate leadership style.

In the May 17, 1993 edition of *Fortune* magazine, several key issues were identified as forces expected to shape the workplace of tomorrow. They included:

- The average company will become (is already!) smaller, employing fewer people.

- The traditional hierarchical organization will give way to a variety of forms, with the network of specialists foremost among these.

- The vertical division of labor will be replaced by the horizontal.

- The paradigm of business will shift from making a product to providing a service.

- Work itself will be re-defined through life long learning, more higher order thinking, and less nine-to-five.[13]

In addition, a Conference Board of Canada study developed an employability skills profile on what employers want in their employees. From this list, it is clear that these skills will require a different degree and style of leadership than have most traditional safety initiatives. These skills include:

Academic Skills

- Writing, reading, computer literacy.
- Oral communication skills.
- Critical thinking ability.

Personal Management Skills

- Positive attitudes and behaviors.
- Responsibility and accountability.
- Adaptability.

Teamwork Skills

- Understand/contribute to goals of the organization.
- Ability to work within the culture of the group.
- Exercise give & take to achieve group results.

- Lead groups and teams, where appropriate.[14]

The business of attempting to define work will be done in the future has been the subject of many books. In Richard Worzel's book, *The Next Twenty Years of Your Life*, Worzel paints a picture of the world of work which poses challenges for safety, especially with respect to the traditional ways in which we have approached incentives and recognition. Says Worzel, "As routine work disappears, whether eliminated by automation or lost to workers in developing countries, routine workers will also vanish. The crucial work—and the crucial workers—that remain will be more entrepreneurial, more creative, more thoughtful than the average worker of the past. They will also be more challenging to manage because they will think for and of themselves; they will decide whether they are receiving as much as they are giving and whether they can demand more and get it. ... To be sustained, creativity cannot be commanded, it can only be encouraged. This point is being lost in some of the uncertainty surrounding management today, and the practice that typifies the current fuzziness of thought is downsizing." [15]

So what do these issues have to do with safety leadership? As I wrote in 1993, "Safety professionals who look inward for solutions to existing and future safety problems are looking in the wrong place. As we've already noted, the world does not revolve around safety. In fact, a considerable number of so-called safety issues are simply reactions to other key issues and challenges that our society has been faced with lately."[16] Successful people and accident prevention strategies need to be leveraged and balanced with progressive, high performance safety recognition and incentive options.

While the rest of the world goes about the process of change at a tremendous pace, our safety initiatives fail to even keep up with the past, let alone keep pace with the present. What about safety being at the forefront of these changes, or even helping shape some of the changes? What does this have to do with leadership? Quite simply, everything. Those who are best able to identify key issues which potentially impact their organizations can best articulate what type of safety performance improvement strategy is needed, can most effectively and expeditiously rally skilled people around the issues, and can maximize available resources aimed at minimizing the potential for loss. At the same time, they will maximize safety performance and be the winners and survivors in the safety leadership challenge.

Again, this leadership role is available for anyone who chooses to accept the challenge, and it is most certainly a challenge. There are individuals and groups out there who feel quite strongly that their views and opinions on safety, and only their views and opinions, are the correct ones. They may be right, but I would suggest that those who show the most principled and convincing leadership in safety will be the ones who will get their agenda on safety accepted.

In *Leadership in Safety Management*, James R. Thomen notes an article by the often-quoted Peter F. Drucker. In the article, Drucker looks at the issue of leadership and calls it, "mundane, unromantic and boring. Its essence is performance." The crux of leadership, according to Drucker, is how a leader achieves goals, or for that matter, is able to articulate goals. According to Drucker, "What distinguishes the leader from the mis-leader are his goals. Whether the compromise he makes with the constraints of reality—which may involve political, economic, financial or people problems—are compatible with his mission and goals or lead away from them determines whether he is an effective leader. And whether he holds fast to a few basic standards (exemplifying them in his own conduct) or whether "standards" for him are what he can get away with, determines whether the leader has followers or only hypocritical time-servers."[17]

Interesting, especially in light of management's role in the internal responsibility system for safety performance. Do most safety management systems request one thing, yet promote another? Do they say "safety is a value around here" but on the other hand say "we have to get on with the job of production at all costs"? Or has management been able to articulate what I call the "macro safety issues" (some refer to them as up-stream safety factors) to the extent that what gets translated into day-to-day, practical activities (or micro safety activities) is a relatively accurate reflection of the values which management sets for organizational safety performance?

And what about leadership styles? Which styles work best? Which do not? In Petersen's *Safety Management: A Human Approach*, he notes that regardless of what type of name you put on it, or whatever current management gurus are calling it, the following have been identified as success factors with respect to the relative effectiveness of leadership (management) style:

- Supervisors with the best records of performance focus their primary attention on the human aspects of their subordinates' problems, and attempt to build effective work groups with high performance goals.

- General rather than close supervision is more associated with high rather than a low level of productivity.

- Genuine interest on the part of a superior in the success and well-being of subordinates has a marked effect on performance.[18]

These leadership issues are of a very specific nature. Now consider how safety issues get addressed throughout business generally. Are those in leadership roles more concerned about furthering their own political (safety) agendas at the expense of safety in general, and workers in particular? Has safety become more of a political issue than a people issue for labor and government? Do most managers have enough of a working knowledge of safety management to play a leadership role? Some would argue that management in general doesn't have enough training or education in safety management to be considered knowledgeable enough to demonstrate effective and informed leadership in safety. The debate continues.

The bottom line: leadership in safety management is an essential part of a complete and comprehensive safety management system. How this leadership gets demonstrated, who exhibits this leadership, and the effectiveness of this leadership will depend upon those who choose to take a leadership role in safety. This role is no longer the exclusive domain of any one group or individual.

Leadership can be exhibited by anyone, at any time. It can be exhibited by supervisors as they conduct safety meetings or work site inspections. It can be exhibited by labor OH&S committee representatives as they lobby for or make proposals for improvements in safety standards. It can be exhibited by government as it shows that there is more to health and safety management than mere compliance with acts and regulations. It can be shown by anyone who chooses to step outside the traditional bounds that have constrained significant improvements in safety performance.

Regardless of what type of leadership role you currently play, value judgements are constantly being made about your leadership style, whether you like it or not. The challenge is to craft your safety management leadership style in order to maximize the full potential and resources of your organization and its people, and to get rid of the paternalistic safety rhetoric which does nothing for the advancement of real safety improvement, but satisfies only short-term, political agendas.

CHAPTER 4

WHERE DO SAFETY INCENTIVE AND RECOGNITION PROGRAMS FIT IN?

THE SAFETY/QUALITY RELATIONSHIP

To accept both the theory and practice of many incentive programs means that you accept the theory and principles of both behavior modification and extrinsic or intrinsic motivation. On this point, many advocates and detractors of incentives part company.

One of the most articulate detractors of the behavioral movement is Alfie Kohn, author *of Punished By Rewards: The Trouble with Gold Stars, Incentive Plans, A's, Praise and Other Bribes*. In Kohn's opinion, and based on his cogent arguments, incentives and other types of reward programs are simply another way of getting people to perform tasks and activities which they naturally resist doing.

Through the organizational adaptation of B.F. Skinner's research and work in operant conditioning, those who feel the behavior of others can be controlled by reinforcement feel that with rewards, selected behaviors are likely to be repeated. Kohn refers to as the do this and you'll get that approach, and describes the American workplace as enormous Skinner boxes with parking lots. From the factory worker laboring for piecework pay to top executives prodded by promises of stock options, from special privileges accorded to Employees of the Month to salespeople working on commission, the recipe always calls for rigid behaviorism.[1]

A fundamental element of incentive approaches is that others, commonly referred to by behaviorists as organisms, can be influenced and made to perform in selected ways through behavior modification. While many behaviorists have been criticized for using the phrase behavior modification and have attempted to devise new catch-phrases for this approach, the reality is it is still behavior modification. You can call a safety audit an assessment or an evaluation, but the functional definition is still the same. A safety audit is a safety audit. Behavior modification is behavior modification.

behaviorism: n. 1. A school of psychology that regards the objective observation of the behavior of organisms (usually by means of automatic recording devices) as the only proper subject for study and that often refuses to postulate any intervening mechanisms between the stimulus and the response. 2. The doctrine that the mind has no separate existence but that statements about the mind and mental states can be analyzed into statements about actual and potential behavior.[2]

But whatever the name, the primary objective has been the same: to recognize, document and attempt to reinforce safe performance, to ensure compliance, to identify at-risk behaviors or unsafe acts, and attempt to change them to the required safe behaviors. The difficult question is, "just what is safe performance?"

Is it no disabling injuries? Is it no accidents at all? Is it doing all the prevention activities right, including exhibiting the correct behaviors? Is it luck? Is it managerial commitment and active participation in your safety management strategies and efforts? Is it a total safety management system approach, or is it simple compliance with rules, standards of behavior, or anything else deemed to be the defined performance expectation? Or is it a combination of all of these things? Like my father once told me: there are two sides to every story, and then somewhere in the middle, there's the real story.

The desire for a system which can contribute to continuous, measurable improvement in specific areas of safety performance has been one of the most sought after, yet one of the most elusive objectives of the industrial safety movement. Some answers to this dilemma may exist in the relationship between the total quality management movement to the occupational health and safety field. The solutions to improving safety performance may be found in both the philosophy and practical application of total quality management tools and techniques, as well as the relationship which some indicate it has to accident prevention, safety management and loss control.

But again, the key is to take all the lofty theory from the textbook to the shop floor, to the boardroom, and all offices in between. As many have found out, this is no easy task. While some TQM slogans exhort people to do it right the first time, the road to effective safety performance, characterized by the application of many TQM principles, is strewn with failures and false starts.

In *The Measurement of Safety Performance*, William E. Tarrants notes that it is highly recommended that any system of safety performance measurement not be related to contests or awards, as has so often been the case. The reason is to avoid the phenomenon known as contest contamination.[3] When the goal of safety is to focus its efforts on rewarding employees only for injury free performance, it's been said that there is potential for the credibility of the measurement system to be brought into question. The primary function of any safety performance measurement system or quality measurement is to provide information that is needed for improved management decision making. Then performance improvement targets can be set and, if so desired, reinforced and rewarded when met.

While contests may not be the name of the game, it is only natural that if individuals or teams are measured, they will invariably want to know by which standards they are being measured, whether they have control over the work which is being measured, and whether they can also control their destiny as far as rewards and recognition are concerned. It is not unreasonable to expect that those individuals or groups which meet or exceed expectations should be given some form of recognition for that achievement.

When I say recognition, I'm not referring to artificially created recognition options that are nothing more than PR, propaganda and photo opportunities for senior executives. I'm talking about opportunities created for sincere celebration, based on mutual goals, common objectives and shared safety performance improvement targets. Again, this is no mean feat and is harder than it sounds. It also involves giving serious consideration to what Alfie Kohn proposes as an alternative to traditional carrot and stick approaches, helping create the conditions for authentic motivation through collaboration, content and choice.[4]

Fundamental to embarking on a program of positive incentives and recognition options for health and safety performance is the recognition that health and safety performance cannot thrive on simple injury rate calculations and assessment, or other accident statistic analysis and examination. This is where traditional accident and injury focused safety supporters and conventional safety management theorists and practitioners part company.

In *The Quality Revolution: Threat or Boon to Safety Professionals*, Edward E. Adams notes the main purpose of the quality organization is to provide products or services that are the most economical, useful, and always meet or exceed customer expectations.[5] Those with responsibility for administering and delivering safety to the organization essentially must focus on a quality product (safety) and tailor it to meet the demands of the customer for that product (management and employees). Success in marketing the product depends on how well the customer likes the product, and whether he or she is prepared to be a repeat customer.

It has been argued that the focus of statistical quality control is on equipment, machinery, work processes, and other things, while the focus for safety is people. However, safety and total quality have fundamental similarities in the concept of measurement, evaluation and feedback for improvement opportunities in the entire safety management system. And a well-designed safety management system can effectively measure the performance of both people and things. As an example, worker health and safety committees have been compared to quality circles, and in fact offer a prime launching pad for potential safety/quality activities with a focus on the people process.

Kansai Electric Power, Japan's second largest utility and winner of the 1984 Deming Prize for excellence in quality, used the concept of quality circles to address specific environmental and safety issues. While past activities of quality control circles centered around safety and health, Kansai was able to extend the gains made from the safety/quality circle arena to other quality issues. In the jargon of TQM aficionados, value was added to the product. By 1984, service interruptions and job-related accidents had decreased, and there had also been a fourfold reduction in annual costs. They also found that employee morale improved, teamwork was enhanced and there was a heightened sensitivity towards the market and the customer, proving the potential for the link between safety and quality efforts.[6]

Just as important is the selection of key measures that determine the success of the safety/quality effort. Rather than simple measures of loss, such as accident or injury statistics, pro-active or performance-based measures are needed to serve as balanced, predictive indicators of safety performance. Just as there is no one single criterion, source or cause of accidents, neither are single measurements of safety performance adequate. Multiple measures are needed which effectively measure the state of safety, or the lack of it. This belief has been advocated by the safety movement for quite some time, but its practical application has not been as evident as the theory would suggest.

The human side of quality is perhaps one of the more important aspects of the safety/quality relationship, specifically from the point of view of involving, training and empowering employees and management in the organization to come up with new and innovative ways to improve safety performance. It has been discovered by those researching the safety/quality relationship that total quality and safety are so similar that safety is simply one dimension of quality. But as we've seen, simply saying or implying workers are empowered may not be a true reflection of reality.

It's been said that "safety is a clear cut barometer of organizational excellence. You cannot have an excellent organization that has a lot of accidents." The difference between those which exhibit performance excellence and those which need improvement can be traced back to a strategy which is defined, mapped out, and measured on a continual, timely basis and one which doesn't have these characteristics. In both quality and safety, companies that are high performers plan it that way.

Experience has shown that the essential elements of a successful quality improvement or safety improvement process are not technical systems. Rather, it is the ability to motivate participation, foster a sense of ownership and tap into the brain power of employees and management to improve all aspects of performance, as well as the people management systems which helps determine performance. It has been said that quality is a belief system. It is a philosophy of continuous improvement. It assumes that people are capable of doing things right the first time. Like quality, safety must be a valued aspect of the corporate culture before true, sustained safety performance improvement can be realized. This provides an opportunity for multiple reward and recognition levels, from the executive to the shop floor.

CAUTION AND INTELLIGENCE NEEDED WITH SELECTED QUALITY APPROACHES

One of the cautions about relating the safety and quality areas too closely is that it may be used for less than honorable reasons as a legitimate reason for downsizing. "Do more with less" is a common demand in the business world of the '90s, but to venture into the total quality movement without a firm understanding of how decisions to downsize may impact the bottom line performance of the organization, as well as safety performance, is pure folly. Safety may even be compromised in the case of a superficial, misguided quality effort.

In a survey of some 55 companies which reported reductions in their workforce, with the average decline registering 13% over the previous 15 months, one of the downsides to downsizing was an increase in workers' compensation claims (the percentage of workers filing such claims rose in the wake of downsizing). The anomaly, explained one of the consultants conducting the survey, resulted in increased injuries sustained by surviving employees, many of them older workers who assume unfamiliar jobs. A company that downsizes and decides that total quality management can help ease the transition and contribute to greater effectiveness may be in for a rude awakening. Total quality management or downsizing may not result in a more effective or safe company. It may just mean that there are less people being less effective or safe.

A jointly sponsored study by Ernst and Young and the American Quality Foundation found that total quality management may not be the panacea which some believe it to be. The potential exists, but only if managed and implemented correctly. The reason, notes the study, is that most organizations can be classified as low, medium or high performers, and need to utilize quite different total quality management tools to improve what they do. Quality in theory is much different than quality in practice. Practical application of safety/quality tools is essential.[7]

One of the areas of emphasis for high performance companies, as identified in the international quality study, was in the area of reliability, responsiveness and safety as key elements for overall reputation. This is not to say that safety cannot become an integral strategic link at companies once identified as low or medium performers. However, the best of the best say that safety, as part of a strategy that focuses on continuous performance improvement, is of prime importance. Safety adds value and enhances competitive position.

The potential for improvement in safety performance, using techniques developed and refined by total quality management movement, lies in the development of specific processes, strategies and measures, and having everyone in the organization understand how to utilize the safety/quality relationship to maximum competitive advantage. Tools such as safety/quality circles, cause-and-effect diagrams, pareto diagrams and various sampling and statistical techniques can all be part of the safety/quality relationship.

In *Quality Is Free: The Art of Making Quality Certain*, Phillip Crosby notes that organizations which consistently perform to high success levels do so because they plan it that way. By using the tools of total quality management, the potential to continually improve all safety processes exists. When I asked him about the safety/quality relationship, his response was that a lot of his writing and focus has been aimed at prevention. In several of his books, he has referred to safety as an analogy for quality because he felt it was useful in explaining the concepts of prevention. According to Crosby, people listen closer when the talk is about how many accidents should be planned to occur, or avoid, each year. [8]

Here is where the potential for improving safety performance lies: the successful integration of appropriate total quality measurement principles, used in the successful improvement of selected safety processes, the measurement of performance-based activities which allow for timely intervention and prevention of accident causing scenarios, and performance improvement.

AVOIDING DYSFUNCTION: DO YOUR HOMEWORK

In the September 1996 issue of *Safety & Health*, an article entitled "Can Safety Be Too Much Fun" suggested that while some safety games motivate employees to work safely, others may do more harm than good. The most interesting point in the article noted that because of all the perceived problems associated with traditional incentive award programs—especially the assertion that employees will not report injuries if a gift or award depends on their injury record—members of an OSHA advisory committee have expressed "strong concerns that safety contests lead to employees not reporting work-related injuries and illnesses ...part of the twist on the argument notes that section 11(c) of the federal OSHA act, and similar provisions in each of the acts administered by the state OSHAs, provide that employees can't be discriminated against for exercising any right under the act. Since employees have the right to report that they have been hurt on the job, any games that discourage the exercise of that right could be considered discriminatory." [9] That's the theory and argument.

So how do you go about deciding if safety incentives and recognition alternatives are right for your organization, and if they are, how they should be structured? How do you come to grips with recognizing and reinforcing good safety performance (presumably you have definitions of good safety performance)? If these approaches are not for you, what does this say about you and your organization?

There has been a great deal of research and discussion on the positive and negative impacts of various approaches to safety incentives over the past 20 years or so. While many organizations continue to use prizes and awards for no lost time injuries, there still has not been any definitive correlation shown between successful safety performance and incentive awards, although organizations which use them infer a connection. In *Safety Management: A Human Approach*, Peterson notes, "organizations use dinners and awards simply because it was traditional for safety programs to use them, but in short, there was never any real foundation for using the element of safety incentives in our safety programs." [10]

Some recent research, however, would suggest that incentive awards" are still popular, but that the prerequisite for getting the rewards has been challenged for change. In *The Psychology of Safety*, E. Scott Geller puts forward a number of arguments about the human condition and human behavior, and the basic purpose of his book, as he acknowledges, is to "explore the human dynamics of occupational health and safety, and show how they can be managed to significantly improve safety performance." In espousing a set of some 50 principles for what Geller terms a Total Safety Culture, he notes in principle number nine that safety incentive programs should focus on the process rather than the outcomes. In other words, you should reward what people do (specific, risk-reducing behaviors), rather that the consequences of what they may or may not have done.

When we examine Geller's approach within the context of the accident theory of multiple causation, specifically as it relates to some of the causes of workplace accidents, the approach seems to makes sense, at least to those who support and advocate the behavior-based approach to safety. Says Geller, "One of the most frequent common-sense mistakes in safety management is in the use of outcome-based incentive programs. Giving rewards for avoiding an injury seems reasonable and logical. But it readily leads to covering up minor injuries and a distorted picture of safety performance. The activator-behavior-consequence contingency demonstrates that safety incentives need to focus on process activities, or safety related behaviors." [11]

This approach was not unlike that advocated by other safety professionals and some quality advocates, although some are advocating taking the rewards higher than work behavior levels. In *Total Quality for Safety & Health Professionals*, F. David Pierce notes, "Historically, we have used safety awards as carrots for worker safety. Most times these focus on workers staying injury free, not on worker safety participation. It's for this reason that these injury-free-based award programs have mixed results. Participation-based awards, such as department awards for high safety participation, are different. They can significantly build employee participation. When used, they can change the perceptions destructive to safety award programs. That is, they bring a halt to the reward systems that depend on not having injuries and, instead, focus on involvement." Pierce notes, "The last tool you can use to build participation is probably the most effective. Tie individual safety accountability to each worker's compensation program, including management and workers. So those who are highly involved in safety and keep the participation high are rewarded with pay increases. Pay is the most effective carrot." [12]

As a point of interest, the November/December 1996 issue of *ACA News* noted that recent studies indicated compensation programs are becoming more strategic. Sandra O'Neal of Towers Perrin noted that research has determined that many respondents are developing or reassessing their total compensation strategies to ensure a closer link with their existing or planned business strategy. Many are also changing their pay strategy to emphasize more variable, performance-based components. This is being done primarily to guide organizational and individual behavior, not simply control costs. Additionally, recognition programs are gaining new attention as tools for rewarding and publicizing employee excellence. Even though the world of work is changing at a rapid pace, the changes to compensation strategies have been deemed to be more evolutionary than revolutionary. [13]

A TRADITION CONTINUES, BUT AT WHAT PRICE?

The practice of rewarding or recognizing employees for safety achievements has been a long-standing tradition in numerous organizations. Its origins are not clear, but most safety people have used them in one form or another. A lot of what we see in the popular literature and textbooks regarding safety incentive awards appears to be US in origin. There is little original research or information on this issue from a Canadian, European, Asian or other international perspective. There is a clear distinction between US culture and many other cultures of the world, and some have suggested this cultural change has the potential to impact safety, and the ways in which we perceive management and worker involvement in safety, including theories of accident causation, systems for prevention, and safety incentives.

While the intentions of recognition or incentive programs are generally honorable, they are often ill-designed and poorly thought out. They are based, to a large degree, on the principles of behavior modification and motivation (with particular emphasis on the antecedent-behavior-consequence model), sometimes by individuals with just a cursory or passing knowledge of behavioral issues.

The primary approach used in behavior modification programs is to first attempt to eliminate unwanted or at risk behaviors that may lead to accidents or injuries, or to create acceptable new responses. While some may consider these principles to be sound, when administered by people with only a fleeting knowledge of how these principles work, they sometimes produce the exact opposite of what they were intended to achieve. Like the saying goes, a little knowledge can be a dangerous thing.

But the real debate centers around how effective these programs are, and whether they really help improve safety in the long run. There are many opinions that both support and shoot down the safety recognition theory, but what real evidence exists, and is it conclusive and absolute? This is where we have to give serious consideration to the academic studies that focus on the theory of human behavior, compared with those workplaces in which these programs have been tried, and proven to be either effective or worthless.

There have been numerous studies conducted on the factors that constitute effective safety management activities. One of these studies was conducted by A. Cohen and M.J. Smith of the National Institute of Occupational Safety and Health, noting:

1. People work more safely when they are involved directly in decision-making processes. They have to be given a channel to communicate their thoughts to management and receive positive feedback.

2. People work more safely when they have specific and reasonable responsibilities, authority, goals and objectives with respect to identifiable safety performance standards.

3. People are more highly motivated and work more safely when they have immediate feedback about their work.

Cohen and Smith's study indicated that among industry leaders in accident free hours, use of monetary incentives was played down, and management frequently expressed the opinion that safety contests, give-away prizes and once-per-year dinners simply did not work.[14]

Barrie Simoneau, the safety director of the Mines Accident Prevention Association of Manitoba, feels that one of the reasons why so may people use safety incentive and recognition programs is because it simply has been traditional to do so. As to why they have been a popular part of traditional safety efforts, Simoneau says, "I think because people didn't and perhaps still don't understand the relationship between safety, production and quality. This is something I have as a philosophy and I use it fairly often: safety performance must be a measure of success, not an analysis of failure. We tend to measure safety performance in terms of loss—frequency, severity, damage—and we make these beautiful charts and put them in annual reports, and we say look how bad we're doing. Then we try to analyze those failures to see what the problem is. Really, what we need to do is rather than deal with symptoms of problems, we need to identify the problem, solve the problem, and you never need to talk about safety."[15]

OTHER REAL WORLD APPLICATIONS

John Blogg is secretary and manager of industrial relations with the Ontario Mining Association. Rather than having safety performance receive special reward attention, Blogg feels safety should be treated just the same as any other part of the job. Says Blogg, "I have a problem with rewards and safety performance because safety is simply doing the job right. If you have proper policies and procedures, and good supervision, you have, as a result of that, good safety performance. As a general rule I think health and safety is part of doing the job right, and nothing more than that. When you separate that and give it special recognition, you are saying, I think, and sending a message to the workers that it (safety) is separate and distinct from doing the job."[16]

There are numerous organizations in the United States and Canada that have structured safety recognition and incentive programs in place. Peter Edmonds, the director of safety engineering with the Canadian Lake Carriers Association, is just one of many safety professionals who feel recognition and reward programs are worthwhile. Edmonds notes that recognition takes place in a number of different ways within his association. Says Edmonds, "Each individual member company within our association does it a little different. We have a couple of organizations that do plaque presentations to the ship and give jackets and all types of other incentives, and for them its working exceptionally well."[17]

Nicholas A. Bartzis, safety officer with OSF Inc., notes, "I personally am opposed to incentive programs as it does not have concessions for substandard acts. If substandard acts are not reported, and the focus is lost time accidents, the impact of the program is questionable. Getting people back to work promptly should not be rewarded. The accident itself should have been prevented in the first place. Safety should not be an item that requires positive reinforcement. The consequences of accidents should be a deterrent to substandard activities. If people are getting hurt, spend the time and energy in areas such as training and development, follow up interviews, ways of preventing accidents."

In Bartizis' opinion of incentive programs usually go two ways:

"1. They have a slow and agonizing death. You will have to drive the program, and when you do not have enough time to administer it, people will forget about it.

2. The reward becomes insufficient. People by nature want more and more. If, because of budget constraints you can no longer give out jackets, what will replace them? And what about the person who has received a jacket? What will they then get? Where will this program be down the road, and what will you give away next?

Gifts and promotions modify behavior, but they motivate behavior in the wrong way. We as professionals should promote safe behavior and not a gift package." [18]

Mark Hitz of Raytheon TI Systems, notes, "At my facility, we use a metrics process called the Oregon Productivity Matrix (OPM) within our business groups. They track a number of items (i.e., quality, cycle time, delivery, environmental costs and safety). Each item is weighted as a percent of the whole. Each item has a scale with an expected goal and an ability to exceed the goal (perfection/greatness/world class). For example, the scale is one to 10 with a goal set somewhere around seven. Safety is integrated into the business team and is naturally tracked as a part of their Business Excellence OPM and is not overlooked."

Hitz continues, "The safety metrics can be whatever is important to maintaining a safe operation (reactive measurement is injuries, proactive is training, audits, team making safety improvements, achieving safety thrusts). They have ownership. It works great. Always look for ways of integrating safety into business processes. The problem with stand-alone safety incentive programs is that they are what they are: programs, not processes. Every year someone will ask if it is worth the money and potentially cut the funds. Their effectiveness is always in question. Plus, it is usually driven by the Safety Office not the employees or management."

Hitz recalls, "Many years ago when I was promoting one of those programs and our goal was an incident rate of 1.0, an employee said "you mean you will pay us if we only let x number of people get hurt this year?" He had a point. Now our vision is zero injuries. To get there we have a number of proactive and reactive metrics with annual percent reduction goals in incident rates and measurable actions to target issues. People will forever debate if putting a cash incentive carrot in front of people makes people not report injuries. During one such debate on having a recordable incident rate metric in our OPM, a senior production manager made a point I will always remember. He said, "If we are concerned about our people not reporting injuries then we have a quality problem." Should we get rid of quality defect goals for fear of employees not reporting defects? No. Rather, we must work to always understand that safety is quality. A defect is a defect. We also do give awards and recognition lunches, but they are tied to other things (safety involvement, method improvements, etc.) rather than just a lack of injuries." [19]

Dave Adams, health and safety coordinator with Fletcher Challenge Forests, notes, "We have used safety incentives in the past in the forest industry and they have worked well. There are some lessons we have learned:

1. Do not have a formal incentive scheme based on set targets with set prizes, as people begin to expect it and if you change it, you have a de-motivating effect.

2. Reward areas of safety performance that will further your safety management (E.g., if near miss reporting is dropping, then reward an individual or group for reporting near misses. Do it once and totally unannounced.)

3. Treat safety incentives like a surprise birthday party. The impact is better.

4. Do not reward with material goods or money. We are instituting more of the warm/fuzzy type of awards. (E.g., Our top tree-faller this year will receive some sponsored chain saw equipment, but our company will pay for him and his partner to stay at a top hotel, all expenses paid.) The material award soon goes, but the memory lingers and the safety message goes home and the self esteem lifts as well.

5. Reward for something exceptional and not for just doing your job. Safety is part of the job and should not be treated as an addition to it." [20]

Mike Duram of Louisiana State University says, "Based on my experience with incentives, change from time to time in the program structure and incentives is necessary to maintain freshness and thus interest. Getting the input from the workers is important, but you will probably find that they will favor money more and more as they accumulate the more popular items, such as jackets, caps and other goodies that are inherently popular. We used an approach that worked well for a number of years in a company that I used to work for, and the approach consisted of issuing certificates each month that the employee went without an injury for which he/she was responsible. The certificates were of a certain point value depending on the relative risk of the work activity, and could be redeemed at any time at a national discount chain store."

"The employee had ultimate choice and was able to take advantage of the full value of the certificate and redeem the certificates at sales. The certificates were processed by the chain store and we were invoiced at the company. As time went on, we developed additional goals that were rewarded using the certificates, and also a safety suggestion award program using the same certificates. Employees could accumulate them over a period of time, even years, and purchase items such as cameras, TV sets, etc. Long before the employees tired of the program, we tired of the administrative activities in the safety section. I found that the companies who offer catalogs of gifts and allow redemptions by mail usually mark up the cost of items to the point where a lot of the budget money was lost. Also, some of the gifts were of low quality when received." [21]

MORE SOPHISTICATED APPROACH NEEDED

The major criticism of most safety recognition programs is that they are mainly reactive, and have as their focus the rewarding of a no-lost-time and accident-free record. Critics of this approach question the rationale of a system based on this focus on injuries. Some have said that the only people who determine whether or not an accident results in an injury with time lost away from work are the workers' compensation systems and physicians. They also say it encourages underreporting or false reporting of injuries, and does nothing to improve safety management or associated systems. Others claim that lost time/accident free records and milestones associated with them and their rewards have nothing remotely to do with improving safety performance. Simply waiting until the end of a calendar year and giving out recognition based on the record is artificial.

In a profession known for its clichés and analogies, some have likened the concept of giving safety awards for lost time accident free records to hitting a baseball. They say if you want to teach a hitter how to hit well, you don't get him to emulate a 100 hitter. You get him to study a perennial all-star, or a career 350 hitter like a Mike Piazza or a Ken Griffey, and you try to get the hitter to do all the right things that these players do. Some professionals say that by using lost time accidents as your reward indicator, you are actually measuring failure rates. That's not good enough.[22]

The important issue, say many safety professionals, is determining what efforts contributed to the record and whether they can be documented and duplicated. Others have suggested that one of the greatest failings of these programs is that they are designed as short term cures for long-term safety issues. Many note that their biggest challenge is finding a program that has staying power.

But Barrie Simoneau has his own thoughts about safety, awareness and behavior. Simoneau feels that safety is nothing more than an attitude. “Everything else from there on in is production work,” notes Simoneau. “Then you have to build quality into that production. If you have quality production, you never need to worry about safety.”[23]

CHAPTER 5

DO INCENTIVES WORK OR NOT?

In the January 1995 issue of *Industrial Safety and Hygiene News*, E. Scott Geller quoted from a book by Alfie Kohn entitled *Punishment by Rewards.* Geller quoted a section of Kohn's book in which Kohn states, "Rewards are not actually solutions at all; they are gimmicks, shortcuts, quick fixes that mask problems and ignore reason.... Giving people rewards ... is an inherently objectionable way of reaching our goals by virtue of its status as a means of controlling others.... What rewards and punishment do is induce compliance." [1]

If, as Kohn suggests, these safety incentives and rewards are a means of controlling people, or are perceived as controlling people, there are potentially many implications for safety, some of which may not be as positive as some may hope. Geller notes that, "What Kohn—and others such as the late W. Edwards Deming and Stephen R. Covey—are saying is this: Common safety tools such as incentives, recognition, praise and penalties do more harm than good in the long run because employees see that these tactics are a means of controlling behavior. Feeling controlled, an employee's own inner motivation suffers." [2]

The work of Alfie Kohn is frequently referenced by those who hold a high degree of scepticism and cynicism towards recognition and incentive programs. In *Punished By Rewards*, Kohn argues that carrots and sticks are ineffective at producing long-lasting attitudinal or behavioral changes in workers, and incentive and recognition programs produce nothing more than temporary compliance. According to Kohn's research, surveys suggest that at least three out of every four US corporations rely on some sort of pay-for-performance program to motivate employees.

But Kohn also looks at some of the past and current research on the issue, and notes that some two dozen studies from the field of social psychology have shown conclusively that people who expect to receive a reward for completing a task simply do not perform as well as those who don't expect to receive anything. Additionally, Kohn asserts that in the workplace, not one controlled study, to the best of his knowledge, has ever demonstrated a long-term improvement in the quality of performance as a result of rewards.[3] Also interesting is the fact that, for the most part, these studies never include any anecdotal input or feedback from the subjects for whom these approaches were developed for in the first place—the workers.

If there is one piece of advice which Geller gives in support of the behavioral-based approach to safety, it is that, "The intent must not be to control people, but to help them control their own behavior for the safety of themselves and others. This is why the terms such as behavior modification, discipline and enforcement are inappropriate. They carry the connotation of outside control.... The bottom line is that behavior is motivated by consequences that are obvious and immediate." [4]

Geller may have a point, but it may not be absolute. In my own case, as an example, I exhibit certain behaviors with my retirement saving habits that, in terms of obvious and immediate consequences, are both unknown to me. On the one hand, I'm not sure how much money I may need or will have 20 years from now, and the benefit to me in the short term is nil. Let me tell you, I'd much rather have some fun with it and spend it now, which would indeed make the consequences of my behavior obvious and immediate.

But then again, I'm not some strictly controlled lab experiment or academic exercise, and I have both short and long-term objectives in mind. Also, it would be interesting to see if the behavioral approach is as beneficial and effective for middle and senior management, especially those who are highly motivated and self-managed, with certain career goals and plans and much more opportunity in their long-term employment prospects.

Thomas Krause and his colleagues, in their book *The Behavior-Based Safety Process*, note that behavior, safe or otherwise, can be interpreted by examining something referred to as ABC analysis (Antecedent—Behavior—Consequence). The antecedent (an event which triggers an objectively observable behavior) results in a consequence of that behavior. If the consequences of a behavior are positive, those behaviors will be reinforced and more than likely be repeated. This is why some suggest that objective, observable behaviors and activities, rather than simple accident or injury statistics, should be the focus of safety recognition, if recognition is to be used at all. Focus on activities and behaviors, not attitudes. Attitudes are neither positive nor negative. Their interpretation is often a matter of opinion, subjectivity and value judgement. Activities and behaviors, on the other hand, can be objectively identified, categorized, observed and measured. [5]

The value of the behavior-based approach to safety, according to behavior-based advocates, is that it gets to the front line workers and attempts to address both at risk attitudes and behaviors. In order to reduce the potential for the final link in the accident causation chain to break down, workers need to be very aware that for whatever reasons they may be taking risks, it is the reduction of those risk taking actions and behaviors which will bring the most success from a risk management, injury prevention perspective.

The behavior-based safety approach is currently very popular. For those who have struggled with the people side of the safety equation, it appears to be offering practical solutions. But of course, a fundamental question which must be asked about any approach to health and safety management is, just exactly what is the problem? If it is that workers are not buying into whatever health and safety management strategy is being offered, and all else has been tried, many have suggested that there's nothing wrong with then attempting to change the at risk behaviors, while at the same time attempting to build a more positive, pro-active safety culture and associated value system.

LABOR VS. BEHAVIOR-BASED SAFETY: NO LOVE LOST

While there has been much written about the benefits of the behavioral-based approach to safety, very little has been written about the potential negative aspects of these approaches. As many of the behavioral-based safety materials are written as marketing material, a reasoned examination of the opposite perspective on behavior-based safety would be valuable.

Nancy Lessen, senior staff advisor for strategy and policy for the Massachusetts Coalition for Occupational Health & Safety, during a July 10, 1997 subcommittee hearing on public health and safety before the Senate Committee on Labor and Human Resources in Washington, DC fired both barrels at the safety incentive and behavioral-based safety camps. The following is from Ms. Lessen's cogent testimony before the Senate committee:

"Currently popular and rapidly spreading are employer policies and programs which place the blame for work-related injury and illness on workers themselves. These programs are problematic because they miss the real problems regarding workplace health and safety, because their use results in skewed data that hide the true nature of injury and illness experience in many workplaces, and because they serve to prevent the identification and correction of workplace hazards."

"'Safety Bingo' and other safety incentive programs are programs that provide rewards to workers who do not report suffering work-related injuries and illnesses. Proponents of these programs say they promote safety by rewarding workers for working safely; our experience informs us that what these programs actually do is promote the non-reporting of work-related injuries and illnesses. Research conducted recently by professors from the University of Puget Sound and the University of Colorado uncovered important findings regarding the use and impact of "safety incentive" programs in the workplace. The researchers had been investigating the relationship between employee ownership (employee stock ownership plans and producer co-operatives) in wood products mills, and workplace safety. In a paper printed in the journal *Economic and Industrial Democracy*, Vol. 17:221-241 in 1996, these researchers state, "[Our] fieldwork alerted us to the salience of issues that we may not have anticipated (i.e., the widespread use of safety incentives in several mills)." Once researchers' interviews with workers confirmed the presence of safety incentive or safety bonus systems in the plants, questions were added to the researchers' survey instrument to gain further insight on the impact of these systems."

"In the same paper, the researchers stated, "Our fieldwork indicated the extensive use of bonus or incentive schemes. These incentive schemes give out rewards that range from dinners or jackets to money based on either the individual or work team going a certain number of consecutive days without a lost-time accident. ... Our interviews with managers and workers clearly indicated that team-based awards often end up creating substantial peer pressure not to miss work. In one ... extreme but still illustrative example of the power of these bonus schemes, a worker who lost the tip of his finger in an industrial accident returned to work from the hospital on the same day saying, "I don't want to lose my $50." The researchers concluded that the validity of the "days of work missed due to a work injury" measure of injury severity was in question and noted that their research results "strongly suggest[ed] serious under reporting of injury experience."

"In 1993 the Central New York Council on Occupational Safety and Health (CNYCOSH), MassCOSH's sister COSH group in upstate New York, adopted a resolution opposing the use of safety incentive and safety bingo-type games in the workplace. Their resolution indicates that a number of employers in the central New York area have adopted safety incentive games as part of their health and safety plans and states:

"Workers lose because these games:

- Encourage workers not to report injuries or illnesses, even if they occur.
- Pressure workers into making light of accidents or injuries for fear of ruining the game for fellow workers.

- May delay the identification and correction of health and safety hazards."

Lessen's testimony also included other example of how employees are pressured to ignore their injuries in order to receive incentives. "Several years ago in southeastern Massachusetts, the personnel director of a small firm visited an employee who was at home suffering from symptoms of carpal tunnel syndrome to remind her that if she did not report this injury as work-related, she would still be in the drawing for the big screen TV."

Lessen described typical safety contests. "All those who do not report a work-related injury or illness for a certain period of time are eligible to be in the drawing. The prize is a shiny new pick-up truck...Everyone who does not report a work-related injury or illness for one year is invited to a big banquet where there is a drawing from the names of the "un-injured". The prize in this case is a check for $10,000."

According to Lessen, "... an insidious twist to these "safety incentive" programs occurs when the element of "peer pressure" is added. An entire department will, for example, be given bingo cards. The game will continue until someone in that department reports a work-related injury or illness. At that time, everyone has to turn in their markers and the game starts over. Imagine the pressure on the poor worker who slices their finger or suffers some type of sprain, to not report their injury because a co-worker is about to reach "BINGO" and win the VCR or microwave oven or romantic weekend trip for two."

Peer pressure can cause underreporting of injuries, as Lessen Illustrated: "...a worker who suffered an industrial accident that knocked him unconscious ...woke up as he was being carried on a stretcher into an ambulance. He attempted to get off the stretcher and told the ambulance attendants that he couldn't go to the hospital because he didn't want to lose his $50 safety bonus."

"Workers can not control the conditions which lead to most work-related injuries and illnesses. They can control whether or not they report an injury or illness. Safety incentive programs manipulate the thing workers can control—the reporting of workplace injuries and illnesses," Lessen reported, adding that "... most disturbing is that their use appears to be growing. At two recent conferences of union leaders and members, one in Massachusetts and one in New York, I asked the audiences of 100-200 participants how many had experiences in their workplaces with "safety incentive" programs such as the ones I've described above. One-third to one-half of the participants raised their hands."

On the theme of behavioral-based safety, Lessen was no less articulate and direct. Noted Lesson, "These programs are marketed to employers by individual consultants as well as large firms, with the claims that employers who focus their attention on and promote "safe behavior" among their employees will have resultant lower injury and accident rates. The theory upon which these programs are based is quite simple: workers' unsafe behaviors are responsible for an employer's injury and accident experience.... In many of these "behavioral safety" programs, workers are trained to "observe" other workers and note down when co-workers "commit" an "unsafe behavior." An on-going process of "rewarding" safe behaviors and, in certain cases and programs, punishing "unsafe behaviors," is instituted."

Lessen explained that, "Behavioral safety programs in the workplace put the focus of attention on individual "worker behavior" rather than management-controlled work systems that lie at the root of the vast majority of workplace injury, illness and death in this nation's workplaces."

Lessen provided an example of a worksite utilizing a behavioral safety approach. "This Boston workplace is an industrial setting where many in this multi-lingual, multi-cultural workforce of approximately 150 workers work with machinery that periodically allows objects to get stuck in the machine. The workers have been well-trained in the company rule that governs what to do when something is caught in a machine: they are to turn off the machine and call maintenance to fix the jam. They are instructed never to place their own hand in the machine to dislodge the object. This training has been provided in three languages, and is reinforced by multilingual signs hanging in the workplace. When periodically workers get their hands caught in the machinery and suffer injuries, they are tagged with the label of "unsafe worker" who has committed an "unsafe act.""

"What is not apparent to an outside observer of this situation," Lessen explained, "is the other set of "rules" that are operating in this workplace...each worker has a production quota. Lack of production on any day or part of one day is counted against the worker and results in discipline up to and including termination. Recent downsizing in the maintenance department of this facility resulted in workers waiting 20 minutes to two hours from the time they call maintenance to the time a maintenance department employee comes to dislodge the object. The lack of production during those 20 minutes to two hours is counted against the worker, and can seriously threaten a worker's job. Workers at this plant understand the "written" rule of what they are supposed to do; they also understand the "real" rules of this plant: if something is caught in your machinery, you better find a way to dislodge it quickly; in other words, you better put your hand in that machine. But, if you happen to get injured in the process, it will be "your fault" for not following the "written" rule of the company. Workers at this plant are caught having to make unconscionable "choices" between their jobs and their health—or their limbs, in this case."

Lessen's critique of behavior-based safety continued with "... their focus on "worker compliance" with the wearing of personal protective equipment.... Using the "hierarchy of controls" that guides good industrial hygiene practice and is spelled out in the Occupational Safety and Health Act, personal protective equipment is the last choice among control measures when it comes to abating health or safety hazards in the workplace. Attention must first be given to eliminating the hazard all together. If that is not possible, engineering controls that prevent worker exposure are the next best option. At the bottom of the list comes the wearing of personal protective equipment. This is because no personal protective equipment is fool-proof...."

"In too many workplace situations," Lessen continued, "employers focus on personal protective equipment as the "only" solution for reducing worker exposure to harmful agents, with training on the use of personal protective equipment being the only health and safety training workers receive."

Lessen provided an example of this: "At the General Electric Plant in Lynn, Massachusetts, the focus of management's safety program was on the wearing of safety glasses. The union health and safety committee representatives in this workplace ... analyzed the company's OSHA 200 Logs of workplace injuries and illnesses and discovered that one-half of one percent of worker injuries involved injuries to the eye. Much more common were back injuries, repetitive strain and other injuries not "solved" by the wearing of safety glasses or any other personal protective equipment.... Management's sole or predominant focus on wearing personal protective equipment and workers committing "unsafe acts" by not wearing their personal protective equipment in this workplace obscured the true nature of what was causing most work-related injuries and illnesses and misdirected attention away from addressing hazardous workplace conditions."

BEWARE OF DYSFUNCTION

One of the most high profile cases of an incentive program gone bad took place in 1992 when Sears Auto Centers in California came under severe criticism for making unnecessary repairs to cars brought in by customers. The Auto Center reward system was based in part on the number of parts service managers sold. The service manager allegedly put pressure on service mechanics to install parts that were not needed. The company was eventually charged with defrauding customers and, reportedly, the settlement by Sears was in the area of $15 million. Incentive programs can, as this example shows, result in some very serious consequences, which are not necessarily positive.[6]

There are any number of ingenious ways managers and employees can come up with to save the record or fix the numbers, none of them having anything to do with improving safety performance. But Thomas Krause notes that any system of improving safety performance should not be strictly numbers-based. As founder of one of the leading behavior-based safety consulting firms, Krause feels that one of the reasons why incentive awards should not be centered on injury rates is that this measure is too unreliable and subject to manipulation, thereby possibly resulting in dysfunction of the award program. Krause and others feel there are six very important reasons why accident data should not be used as the primary indicator of safety performance:

1. This approach is reactive rather than proactive.

2. Random variability is misread.

3. As a consequence of random variability and its negative effect, management overreacts.

4. Safety incentives based on frequency rates amount to false feedback. You may actually be rewarding unsafe behaviors which, through luck, happen to result in a low frequency or severity rate.

5. The emphasis on frequency rates encourage mere numbers management, not improvements in policies, procedures, training or behavior expectations.

6. The net result of the first five factors is an erosion of the credibility of the safety effort. [7]

The focus, say those who support the behavior-based safety approach, should be on day-to-day actions of people, not simply yearly highlights which may have been achieved through luck. More specifically, those day-to-day activities can be narrowed down to measurable behaviors In other words, behavior modification forces management, supervisors and employees to focus on specific areas that need improvement. With a behavioral approach to safety, the organization:

- Defines what types of behaviors are necessary for safe performance.

- Trains employees in safe behavior.
- Establishes a system to observe proper behaviors.
- Uses positive reinforcement and feedback when an employee displays those behaviors. [8]

The recent shift in thinking on how to structure a performance-based safety recognition or incentive program has been due in part to the relationship between safety and total quality management concepts. Jargon such as process, measuring performance and continuous improvement have become just as important a part of the modern safety vernacular as frequency and severity used to be. Yet the emphasis on what some are calling the process seems to be creating more opportunities to not only modify and influence safe behaviors, but to recognize them as well.

It's been suggested that one of the strengths of a TQM approach over a behavioral approach to safety is that TQM, or whatever the latest buzzword is for this improvement process, focuses on systematic changes in attitude, which, in turn, result in changes in behavior. The quality approach has its focus on safety processes, including the safety systems developed by an organization directed towards accident prevention. Some suggest a TQM approach to safety offers more long-lasting results, but behavior modification gives quicker impact, especially with specific, observable problems, as represented through selected, at risk behaviors.

It's been argued that the strength of the quality or effectiveness approach over the behavior-based safety approach is that it emphasizes employee participation, uses internal controls that reinforce both attitude and behavior change, and has the capacity to sustain systemic safety improvement. The weakness of behavior modification, some suggest, is that it focuses on specific worker behaviors only, which may cause a company to ignore the need for emphasis on its safety management system.[9]

BUILDING A SOLID FOUNDATION

The challenges to conventional safety thinking, coupled with the advent of pressure in the workplace to be more efficient and competitive, have forced many in the safety field to evaluate just how effective or useful these recognition and incentive programs are. It is not a simple, black and white issue. They apparently work for some. Others detest them. Yet despite the controversy, incentives and recognition may still play an important part in your safety management process.

It's important to take the time to seriously ask, "Just what is it we want this program to do for us, and how can it best be done?" While the decision to use or not to use safety incentives or recognition is one which only can be made by you and your organization, if you should decide they are for you, there are some guidelines which should help make your decision the right one. In a 1989 study of safety program evaluations, researchers McAfee and Winn explained that specific rewarded behaviors may improve, while other safe behaviors may deteriorate. In contrast, if avoiding accidents is rewarded, then all the behaviors that contribute to it are also rewarded and maintained. The best solution seems to be a balance of both activities and statistical results.[10]

There are some common features to all behavior-based approaches to safety which need to be considered before attempting to initiate these approaches and complementary incentive or recognition options. These include:

1. The belief that worker behavior is the precursor to safety or injury.

2. The implementation must be achieved through training.

3. High participation is critical for success.

4. Management commitment to the process is essential.

5. Behavior is objective and can be observed.

6. Unsafe or at-risk behavior can be objectively measured.

7. Improving safe behavior and minimizing at-risk behavior reduces injuries.[11]

Additionally, more specific approaches which attempt to tie behavior-based safety to safety incentives and recognition recommend the following:

1. The behaviors required to achieve a safety award should be specified and perceived as achievable by the participants.

2. Everyone who meets the behavioral criteria should be rewarded.

3. It's better for many participants to receive small awards than for one person to receive a big award.

4. The rewards should be displayed and represent safety achievement. Coffee mugs, hats, shirts, sweaters, blankets or jackets with a safety message are preferable to rewards that will be hidden, used or spent.

5. Contests should not reward one group at the expense of another.

6. Groups should not be penalized or lose their rewards for failure by an individual.

7. Progress toward achieving a safety reward should be systematically monitored and publicly posted for all participants.[12]

It's interesting to note that this type of approach will work in an organization which has adopted a behavior-based approach to safety, where the organization has developed an inventory of at risk or unsafe behaviors, and samples the workplace to determine the percentage of safe behaviors observed on any given shift, day or week. Research also suggests that the behaviorist approach to safety seems to have its roots deeply planted in the US, with the work of a number of American psychologists apparently leading the way in taking advantage of the opportunity to apply their trade to industrial safety.

A number of safety professionals queried indicate somewhat of a discomfort with the term 'behavior approach,' as it is sometimes perceived as a controlling approach, based on psychological attempts to manipulate human behavior. Also, a number of labor groups queried have difficulty with the behavior modification concept. The cogent argument of Nancy Lessen more clearly expresses this anti-behavioral perspective. Many consider this American approach to be totally inconsistent with the issues of the right to participate, the right to refuse and the right to know, fundamental principles of much Canadian and Scandinavian OH&S legislation.

WHAT DOES THE FUTURE HOLD?

What will the future bring for the role of workplace recognition and incentive programs? What about the whole area of behavior modification and its place in safety? John Blogg, of the Ontario Mining Association, feels these programs will change as the nature and complexion of the workplace changes. Says Blogg, "I think as companies become more sophisticated through technology, they see less opportunity for incentives because technology is taking over, and you don't have as many people involved. I think the work of the behaviorists, and all of the books we have now on health and safety, is certainly leading away from that."[13]

Do incentive and recognition programs produce results, and improve safety performance? When Syncrude's Jim Williams was asked this question, he replied, "I can't answer that.... When I came here in 1977, there was no recognition. Now you're recognized not only for what you do but for who you are."[14]

Nowhere in any legislation will you see suggestions or laws aimed specifically at attempting to influence or modify individual or group behaviors. It may be implied that through training, etc., certain behaviors are expected, but traditional approaches to safety have a number of drawbacks. In his article on "The Behavioral Approach," Michael Gilmore, senior partner with Safety Performance Solutions, notes, "Training alone does not always result in compliance. Just because people know how to do something doesn't mean that they will automatically do it that way—especially if the required behavior seems time-consuming or unattractive. Second, the reliance on supervision as a primary motivator is often at odds with a worker's desire for self-reliance and independence. Compliance based on supervision tends to function only when supervision is actually present. This type of traditional top-down, directive approach to safety can squelch employee empowerment by reducing perceptions of self-effectiveness, personal control and optimism."[15]

The world economic crisis which is forcing most organizations to objectively evaluate the logic and rationale of all programs may be a cloud with a silver lining for those wishing to re-evaluate the expectations for their recognition or incentive programs, especially where there are substantial operating costs associated with the programs. Perhaps it may be the opportunity, in the words of the TQM gurus, to do the right things, right.

Can a balance be stuck between attempts by employers and employees to comply with health and safety regulations, and the ideas and strategies of those who see behaviors as key to preventing accidents? Will legislation ever keep up with the innovations which some employees and their employers are making towards improved accident reduction and human performance? The answer is for employers and employees to come to agreement on the best approaches that they can all equally participate in for the maximum benefits. With all due respect to legislation and its honorable intentions, that's the reality of the workplace.

CHAPTER 6

CULTURES AND EMPOWERMENT

CULTURE IS IMPORTANT, BUT KEEP YOUR PERSPECTIVE

Just what is a safety culture? For that matter, what is a corporate culture, and who defines it? Lately, a lot of people have been talking about safety having to be an important part of the corporate culture, or that safety needs to be a value in the culture of the organization. Despite all that has been written, there is still a fair deal of cynicism associated with culture, especially when it is mentioned as a new cure-all for safety. A considerable amount of the research and practical application of corporate culture is perhaps worthwhile, or at least makes interesting reading. But the current proliferation of corporate culture rhetoric and vernacular may be overkill.

Most of us who work in the safety field try to look for new (and sometimes exciting) ways of keeping safety issues vibrant and relevant in the day-to-day operation of whatever business we're working for. Everyone seems to drowning in the latest management hype, buzzwords, clichés, one-liners and sound bites. In the quest to come up with new and exciting ways to improve safety performance, we sometimes forget one very important thing—the basics. It's important to achieve a practical blend of both theory and reality: the real world, so to speak.

In *Bringing Out the Best In People*, Audrey Daniels notes, "Business professors and consultants have succeeded in making corporate culture change into a complicated and expensive process. The concepts and activities associated with typical attempts to change the culture are not based on solid research and are not implemented in a way that demonstrates cause-and-effect relationships between what was done and what was achieved." [1]

While recognized as important, even if somewhat nebulous, it's been said that corporate culture drives and controls the success of any recognition system that a business may be considering implementing. And while the value of the reward approach may not be spoken about every day, the culture within which these approaches function speaks loud and clear. [2]

William E. Conway, author of the book *The Quality Secret: The Right Way To Manage*, notes that every organization has a different culture, a different way of working.[3] Conway and countless others have noted that is not necessarily the programs or technical aspects of business which ensure its success, but the culture or organizational climate in which these programs exist. In other words, you can have the best safety program, technically, on paper, but if it is forced to operate in a dysfunctional culture, its chances of success are limited.

WHOSE BEHAVIOR: MANAGEMENT, WORKERS OR BOTH?

Krause, Hidley and Hodson, in *The Behavior-Based Safety Process* devote considerable attention to culture. Indeed, the sub-title of their book is *Managing Involvement for an Injury-Free Culture*. They note that accidents and incidents are endpoint, or downstream events, as opposed to upstream events. Accidents and injuries are only symptoms of a safety or prevention process that has gone wrong. The genesis of that problem happens upstream, as a result of the culture which the organization has defined for itself, or as is more often the case, the culture which has been defined by default. Culture dictates your management system and style, and as such, the activities associated with that style. This then determines workplace behaviors and conditions, and thus the potential for high or low accident or incident rates.[4]

Dan Petersen's *Safety Management: A Human Approach*, talks about the value of a positive corporate culture on safety performance. Petersen and others have come to understand that the simple existence of a safety program does not necessarily guarantee safety success because there is no one right or wrong program. What works for you may not work for others. Off-the-shelf, canned safety programs and packages marketed by companies in the business of selling safety may not have all the solutions to your problems, regardless of what the slick, full color brochures may say. By way of example, Petersen cites a study performed by the Association of American Railroads of all major US railroad safety programs. The study found that basically there were no real standard programs. What worked for one did not necessarily work for another. In other words, each organization apparently had its own culture, which dictated the type and degree of safety system that it instituted.[5]

KEYS TO CULTURAL SUCCESS

One of the most important aspects of the cultural approach is that you must first have an honest and straightforward appraisal of your management style, your safety philosophy, and how that philosophy gets translated into day-to-day activities. There are too many safety systems out there which say all kinds of things on paper and in manuals, but practice something completely different. Your safety culture should match all other parts of your corporate culture.

According to some researchers, there are five basic elements that shape corporate culture. They are:

1. Business environment: The reality of the marketplace in which that business operates.

2. Values: What is it that we stand for and believe in around here?

3. Heroes: Who are the role models and leaders, and what are the values and beliefs that they exemplify?

4. Rites and rituals: What are the systematic activities and programs that shape this organization?

5. Cultural network: How the organization communicates, both formally and informally .[6]

In Search of Excellence provides an overview of those attributes which, in the opinion of authors Tom Peters and Robert Waterman, characterize excellent companies and form the basis of a successful culture. These characteristics are:

1. A bias for action: Get on with the job.

2. Close to the customer: Learn from the people you serve.

3. Autonomy and entrepreneurship: Encourage people to excel. Do it—don't just put it on paper.

4. Productivity through people: People are treated with respect. They are the source of excellent performance.

5. Hands-on, value driven: Stay close to the action (the real world) and live the values espoused by the organization.

6. Stick to the knitting: Do what you know best.

7. Simple form, lean staff: If you don't know what to do from day-to-day, what are you doing here?

8. Simultaneous loose-tight properties: Being flexible enough to be both centralized and decentralized, and to know the differences and importance of each.[7]

CULTURE IN THE REAL WORLD

What does all this have to do with the legal requirements for health and safety? Even more interesting, how does labor relate to all these management theories?

Many safety people do not have direct input into what may or may not constitute a corporate culture. Some simply operate on the periphery of the business. However, successful safety pros have learned how to integrate and merge their ideas for a safety culture with other corporate culture issues. The important thing to remember is that "safety is not your number one job," nor can we "keep safety uppermost at all time." We can, however, help to create safety as a corporate value which, when expressed in terms of operational activities, should get just as much respect and consideration as other operational areas.

Safety need not be your number one job, but it should at least be a first among equals. Priorities change. One day safety may be lower on the totem pole than some other issues, but at least if it is a value held throughout the organization, it will not lose its place in the overall scheme of core corporate values. The key to having this value become real and action oriented is to ensure that aside from the soft culture issues, an actual performance oriented safety management system is developed.

Is the issue of corporate culture of any importance to government health and safety regulators? Perhaps not yet. But perhaps it should be. Regulations should be drafted and revised to reflect the changing realities of the workplace. While most regulators feel more comfortable with physical conditions in the workplace and compliance with physical condition requirements, they may not be as comfortable in the areas which some would call soft safety management. Perhaps that's not too surprising, since most governments aren't exactly on the cutting edge of bureaucracy busting, leadership or value setting.

But what if government regulators required in their safety legislation that all companies had to have a written and documented safety system, with the core set of safety values articulated in that document? And what if government worked with industry to promote more excellence in safety management?

Consider the comments of former US Secretary of Labor, Robert B. Reich, in an interview in the National Safety Council's magazine, *Safety and Health*. Said Reich, "The Department of Labor supports any efforts that employers and employees undertake to reduce workplace hazards. The Comprehensive Occupational Safety and Health Reform Act bills require safety and health programs in all workplaces and safety and health committees at companies with 11 or more employees.... Top priorities in standard setting include generic standards for ergonomic hazards, health surveillance that will compliment the rule on air contaminants and mandatory health and safety programs for employers." [8]

Consider as well the comments of OSHA Policy Director John Moran in the March 1996 issue of *Industrial Safety & Hygiene News*. Said Moran, "I want to focus on issues that really have a long-term impact on safety culture in corporate America. This is really challenging. The way we're going to really change things is to make safety an everyday component of life. Regulations alone are easy to comply with. You can spend $100.00 on software and be in compliance. Compliance alone means nothing. The safety and health program standard emphasizes other aspects, like hazard analysis and involving the workforce in audits and so on. Overall, the idea is to turn safety into an important, essential element of doing business. We're focusing on cultural aspects now." [9]

But even in that same interview, Moran acknowledged that the use of positive incentives for employers who are trying to play the safety game, and are achieving good results, also need to be a part of any legislative strategy for improving occupational health and safety. Said Moran, "There are a wide range of options, like the number of hazards identified and abated through programs like Maine 200 is just one. We need to identify companies with the worst workers' compensation records, identifying the most serious hazards in different industries. We need to focus on employers not doing the job. We must maintain our very, very important enforcement role to make sure those few employers who ignore safety and health get dealt with harshly. And we also need incentives for employers who are trying with good safety and health efforts. There is a very definite need for balance." [10]

If governments plan on legislating safety programs, you can bet your next paycheck that it won't be too long before they will start to feel more comfortable specifying the elements of those programs, including issues surrounding culture and values.

BEWARE THE STEEL HAND IN THE VELVET GLOVE

But it's the careful, intelligent and ethical use of the new people management tools that will eventually determine whether the current fascination with the issue of corporate culture will provide the level of safety performance improvement desired. So where does all this talk of corporate culture leave organized labor, which traditionally has seen activities in this area as just another attempt to manipulate unions?

One of the most investigative and informed treatments of this issue comes from Robert Howard's book, *Brave New Workplace—America' Corporate Utopias: How they Create Inequalities and Social Conflict in Our Working Lives*. In his chapter on "Crafting the Corporate Self," Howard notes that formal company programs such as quality circles, employee participation teams, quality of work life projects and the like are expanding through industry. They seem to part of a more general trend: to meet the crisis of scientific management by concentrating on worker motivation and morale and, in this way, win employee commitment to and participation in the corporation's goals for working life. This tends to personalize corporate relationships, and define shared valued and beliefs as part of the corporate culture.

Howard notes that when managerial control becomes personalized, through the values exemplified in the corporate culture, the relationship of workers to the corporation is understood purely in psychological and individual terms. Power and control then become therapeutic feelings rather than actions. As a result, the reality of everyday working life conflicts become personalized and dismissed as personality differences, or worse, corporate social deviance, rather than legitimate subjects of social or political significance: health and safety, for example.[11]

In other words, if we, the company, are willing to go out of our way to create the perception that we are all one big, happy family in this business, then any disagreements, must obviously be your fault. Some have referred to businesses that have purposely created and crafted this corporate culture as cults, not cultures.

Howard is of the opinion that unionism is under attack from the "enchanted corporation that promises to provide for workers' needs in the brave new workplace by replacing the bureaucratic rules and regulations of the union contract with the good feelings, high commitment and trust of the corporate culture." [12] Interesting point of view.

In *The Witch Doctors: Making Sense of the Management Gurus*, Micklethwait and Woolridge describe corporate culture as a type of glue holding the modern corporation together. They note, "Core competencies, renewal, networking, and entrepreneurialism: a cynic might say that this sounds like a fairly vague list around which to build a company. And the cynic would be right.... All that holds it together is its culture—that intangible thing that inspires employees to be self-disciplined and allows managers and their workers (to use two outdated terms) to trust each other. In fact, culture is not as ephemeral as it sounds.... In most cases, the company's culture cannot be captured in some anodyne mission statement.... Corporate culture is not something that just appears. Companies—and bosses in particular—have to work at it." [13]

So how's your corporate culture? Have you even defined your corporate culture, and have employees been part of the definition process? If you want to improve your safety performance, culture definition is an important step in building the foundation for your safety management plan. But it has to be done with honesty, integrity, and with input (and buy-in) from all segments of your operation. A top-down definition and implementation of a corporate culture is just as ineffective as a top-down implementation of a safety management action plan.

Go back to the basics. Get the input of your people. Determine just what it is you want to do, who is going to do it, and how is it going to get done. Then hold all levels accountable for their activities and safety and accident performance. Give people some credit and let them help you in the development of a system which meets not only their needs, as part of their defined culture, but yours as well.

GETTING BEYOND THE EMPOWERMENT RHETORIC: MEANINGFUL INCENTIVE AND RECOGNITION CONSIDERATIONS

Survey after survey has made one thing abundantly clear. Today's workplace is not what it was even five short years ago. Buzz phrases abound about doing more with less, the leaner, meaner organization, re-engineering your business processes, and my personal favourite, working smarter, not harder. Personally, I'm having tremendous difficulty working smarter, but I can guarantee that I'm working harder than I was this time last year.

All over North America, companies, governments, unions and individuals are struggling to try and come to grips with the new reality of work life. After nearly a decade of frantic cost-cutting, "rightsizing" and rationalization, it is now acknowledged that the downside of downsizing is beginning to take its toll. According to Bernard Wyscocki, Jr. of the *Wall Street Journal*, in his article, "The Danger of Stretching Too Far," decimated sales staff turn in lousy numbers. "Survivor syndrome" takes hold, and overburdened staffers just go through the motions of working, worker productivity doesn't improve, and earnings don't go up. Some have coined the phrase "corporate anorexia" to describe the recent round of shedding of people in the workforce—you may get thinner, but you're not necessarily healthier.[14]

While reengineering may have been a boom to many consultants, it was a death knell to many workers who were simply reengineered out of a job. In *The Reengineering Revolution*, Michael Hammer and Steven A. Stanton identify what they call the five is as a standard repertoire of techniques for overcoming resistance: Incentives, Information, Intervention, Indoctrination and Involvement.[15] While Hammer and Stanton never had safety incentives in mind when they wrote their book, their commentary on incentives is nonetheless relevant to our discussion. They note that while incentives are one of the first change-management techniques many people think of, there are limits to their use. In their example, "you can only use the threat of termination with people that you can actually afford to lose. More subtly, incentives are only truly effective when you use them on people whose resistance is motivated by the perception of a tangible loss. If someone is losing authority or income, there are ways of making that up to him. But attempting to bribe someone who is fearful of a new environment, or whose self-image is undergoing a drastic shift, is like trying to reason with a snowstorm." [16]

So what are we to do to try and improve industrial safety performance and help ensure that everyone in the business, as much as reasonably possible, is committed to the same safety goals and objectives, and are all pointed in the same direction? The logic of mutual goals may very well be a path to mutual gains, especially in the reduction of accidents, illnesses, compensation costs and all the other negatives associated with poor safety performance. An answer may lie in the knowledge and skills that all levels of employees in the organization bring to the safety puzzle. This sharing of ideas, knowledge and skills has come to take on a name all of its own—empowerment. But aside from the fancy name and all the other buzzwords that go along with it, employee ownership of select safety efforts has tremendous potential. Here's how it might work.

BROTHER, CAN YOU SPARE A PARADIGM?

One of the first steps in moving to a safety system which truly values employee input and ownership, and empowers workers to take a proactive stand on health and safety issues, is to change our traditional notions about health and safety roles. It's been suggested that the roadblocks to an empowered workforce from a safety perspective come down to misunderstanding, paradigms, fears and corporate culture.

In the book *Total Quality for Safety & Health Professionals*, F. David Pierce notes, "Paradigms work as information filters. Information that agrees with the paradigm is easily accepted, but information that does not is usually rejected and, in some cases, not seen at all. ...

Paradigms therefore have three effects. First, they mask recognition. Second, paradigms limit learning. And third, largely due to the first two effects, paradigms maintain the status quo." [17]

To get an idea about how employee empowerment might be either facilitated or impeded within your organization, look at some conventional and non-traditional approaches to safety in North America and benchmark your organization based on these comparisons. Traditional safety systems are characterized by differences in perception of how safety works, and the roles which employees, management and safety professionals play in those traditional safety systems.

Consider for a minute the safety perceptions of senior and middle management compared to those of the shop floor worker. Any of you who have sat down with senior and middle management and talked about safety quickly realized that value judgements, subjectivity, complaints about employee attitudes, behavior, carelessness and complacency abound for this group. They identify issues, are confounded about how simple safety should be to achieve, and then wonder why people do the things they do. Great rhetoric, but poor problem solving.

Want to see a 180-degree perception shift? Look at your safety problems and statistics and ask the average employee what they think the problem is. Chances are he or she will point the finger at issues which management is (or is not) addressing, and how they are addressing them. Different paradigms, different perspectives: no common goals, objectives or missions, no common tools, no common acceptance on problems and opportunities, hence, dysfunction, disagreement and everybody going everywhere, doing everything. For those of you familiar with the humor of Monty Python, it resembles the 500-meter dash for people with no general sense of direction!

So how does empowerment happen? Who drives the process? Is your entire business, including employees and management, committed to employee empowerment in theory but not in practice? If they are committed, what tangible efforts demonstrate this commitment?

Let's look at employee empowerment and safety from two different points of view: the TQM perspective and the perspective of existing joint health and safety committees, which some have suggested are prime launching pads for employee empowerment efforts in safety.

As noted earlier, one of the most articulate critics of the behavioral movement is Alfie Kohn. Kohn argues rewards and punishment do nothing more than induce simple compliance,[18] with nothing but short-term modifications in the behavior being manipulated or targeted. Lasting improvement, according to Kohn, can only come about by making the workplace more democratic, and giving workers a true and meaningful say in what they do, and how they do it. Through his alternative Collaboration, Content and Choice model, Kohn argues for a shift from the behavioral modification models, complete with their punishment and reward techniques.

Indeed, collaboration may very well be the next issue in the empowerment and teamwork debate. While teamwork is on the lips of many in management, it may not be as practical a solution as many think it is. You might be able to hold individuals accountable for their specific actions and efforts from an accident prevention perspective, but teams can sometimes blur the lines between individual ownership of safety initiatives and collective abdication of responsibility for safety.

CONVENTIONAL SAFETY STRATEGIES AND STRUCTURES: FAST BECOMING OBSOLETE?

According to an article in the June 1994 *Industrial Safety & Hygiene News*, John Wesley, a safety professional with the Coors Shenandoah brewery, now considers himself to be more of a coach than a traditional safety man. What's also interesting is that Wesley was once an inspector with the US Occupational Safety and Health Administration. At the Coors brewery, Wesley coaches 37 safety teams and works as a volunteer safety goal champion to coordinate activity. The focus is on safe behaviors, and the emphasis is on keeping safety in a positive light. "None of this beating people over the head.... It was tried before I came here and it doesn't work." Interesting that a former government man, whose job it was to carry a big enforcement and compliance stick for OSHA, now plainly sees that you can't force compliance, nor can you force teamwork.[19]

As mentioned, the opportunity for existing joint health and safety committees to empower employees from a health and safety perspective should not be overlooked. In the July 1995 *Safety and Health* magazine, S. Pressly Corev of Sonoco Products noted, "Joint safety committees are an excellent way to empower plant personnel by giving them the ability to make a difference in their jobs. Empowerment in the area of safety has positive effects on morale, productivity, service and quality."[20]

Some would argue that a successful health and safety committee may not ensure good safety performance. They would suggest there is perhaps no clear, direct cause and effect relationship here, and those folks may be right. However, with committee rights being guaranteed in Canadian legislation, and with more efforts made on provincial fronts to more deeply entrench joint committee roles and activities in workplace safety, committee members will be asked to play a greater role in the future, whether safety professionals like it or not.

ADAPT OR ELSE: FLEXIBILITY IS THE KEY

F. David Pierce, in his book *Total Quality for Safety and Health Professionals*, notes, "In an empowering environment, employees are given their heads. It releases their creativity and knowledge to run as fast as they can, in whatever direction they choose within the described path. The described path is set by the organization's shared vision and mission." [21]

There is a difference between the traditional, legislatively mandated joint health and safety committees and empowered safety teams that can include the membership of a joint health and safety committee. Conventional safety committees simply comply. Empowered committees and safety teams actually have clear mandates, roles which involve them in numerous safety processes, and they work on solutions to problems, not simply documenting a litany of problems for management or the safety man to solve.

SAFETY CAN ONLY HAPPEN THROUGH PEOPLE

In the September 18, 1995 issue of *Fortune* magazine, the 11th annual quality report noted, "The story speaks volumes about the relationship of workers and management and the value of teamwork in today's global economy. In the flattened network organization, styles of command and control have changed. Collaboration is in. Today's managers and workers alike have to practice co-operation and collaboration with everybody.... The environment increasingly encourages this by devolution of power and delegation of duties, right down to the empowered, self-managed worker." [22]

That same article also makes the following observation on empowerment: "The essence of the idea is simple. Organize employees into teams that can cut across old boundaries. Train them. Put them into jobs that challenge their abilities. Give them the information they need. Tell them what they need to accomplish. Then turn them loose. Self-directed teams make decisions, set their own goals, and take responsibility…."

Now consider this from a safety perspective. How does this challenge our conventional thinking about not only safety and safety initiatives, but also the role and tasks of safety professionals and their staff? And what are some of the impediments to an empowered workforce from a health and safety point of view? Let's examine some realities.

A considerable number or businesses that have safety staff are still struggling to define a clear and acceptable role for those staff. Let's consider a conventional safety program. Rules and regulations are still the responsibility of a head office safety man, and enforcement is also an expected role of this individual. It is essentially a position of perceived power, including decision-making affecting the health and safety initiatives of the organization.

The reality is that this is not an individual or position of power at all. Why? For the simple fact that the 'safety man' can't be all things to all people, not can he or she be everywhere at the same time. As a result, there is limited commitment or ownership from the line, or from employees. Token compliance is offered, and very often, as soon as the safety person disappears, so too does safety compliance.

While the safety person may for a time bask in this limited perception of power, it is actually the most powerless position in the organization. Just as quickly as the safety person is to take the credit for improved safety performance, they had better be just as prepared to take responsibility when safety performance doesn't improve. This is the point where most safety personnel jump off the bandwagon.

Consider the differences between a traditional and modern safety program, as described by Dan Petersen:

Traditional safety program characteristics:

- A central, head office individual with responsibility for corporate safety.
- A team of safety specialist on staff, or one local staff safety specialist.
- Safety is deemed to be a staff function with a manager responsible for safety and a supporting staff reporting to that manager.
- Rules, standards and regulations, developed and prescribed by management.
- Little line commitment, ownership, responsibility or accountability for safety, and minimal involvement in accident prevention or safety management issues.

Characteristics of more progressive approaches to safety:

- Workers and work teams take responsibility for accident prevention initiatives and safety performance improvements.
- Staff safety specialist employed as a safety coordinator and support function.
- Increased emphasis on total communication with all stakeholders.

- Executive and senior management provide strategic leadership for safety strategy. Content and initiatives left to workers and work teams.
- Total integration of safety performance into all job responsibilities and accountabilities.[23]

Having looked at some of these differences, and the potential which each may or may not have to assist in improving safety performance, which style best reflects you and your business? Which one would you rather be part of? Which style best reflects your current business reality?

EMPOWERMENT NEEDS ATTENTION TO DETAIL

In their book, *Empowerment in Organizations: How to Spark Exceptional Performance*, Judith Vogt and Kenneth Murrell note, "The conceptualization and implementation of empowerment points clearly to the need for a new style of effective leadership." The empowerment efforts need to be defined, and their impact on the organization articulated. Vogt and Murrell also describe a number of principles for enacting empowerment, which include:

1. Principle of Congruence: The ways in which empowerment is implemented must be congruent with one another.

2. Principle of Interdependence: It must be clearly understood that when something happens in one area, it will affect other areas.

3. Principle of Excellence: Is the organization ready, willing and able to move forward on the issue of empowerment? If so, is everyone committed to being

the best they can be through the use of empowerment principles?

4. Principle of Integration: Careful planning, understanding of the strategic direction of change, and assessments of current situation are necessary in order to ensure all empowerment processes are clearly integrated.

5. Principle of Process and Direction: A clear, shared vision must be developed and clearly communicated. You must be clear on what you want to do, why you want to do it, and how you will accomplish it.

6. Principle of Coordination: If the plan for coordinating and communicating the empowerment plan is flawed, so too will be the implementation of any empowerment strategy.

7. Principle of Investment: The business must be willing to invest the time and resources necessary to achieve its strategic empowerment goals. Changes to values, beliefs and culture must be supported by the empowerment effort. [24]

Just because you or others in your business may have read the latest book or article on empowerment and feel it's right for you may not necessarily mean it is. An honest appraisal of your existing corporate climate (popularly known as corporate culture), and especially your existing safety culture, will give you a good indication of how empowerment may or may not work for you. Some have been able to implement empowerment strategies to a greater extent than others. However, some employee ownership of safety and health initiatives will pay dividends for you and your organization, depending on how specifically you target those safety and health tasks with existing personnel and resources.

WELCOME TO THE REAL WORLD!

Empowerment may not be a new idea, nor is it the answer to for all things and all people. But with a successfully and strategically managed empowerment effort, employee contributions and buy-in to health and safety improvement initiatives can go a long way to the eventual success of your health and safety efforts. Here is where the theory parts company with reality. Some have suggested that perception is reality, and others have gone one step further to suggest that in their perception, safety professionals and their ideas are out of touch with reality.

Why does health and safety empowerment sometimes fail? Perhaps for the same reasons that empowerment has failed generally. The promise of empowerment is perhaps too unrealistic. Sure, there are success stories, but by and large, empowerment often fails to live up to its own rhetoric, perhaps because management could not come to grips with the reality behind the rhetoric and the need for a transfer of power to take place. While asking employees to assume more responsibility but not more power, we maintain the status quo, the same traditional autocratic management style.

While there are as many gurus hawking silver bullet solutions to all the ills from which industry is suffering, government is still taking a strong stand on due diligence. No soft sell or modern management theory for those folks: comply with the legislation or suffer the consequences. Some have suggested that the direction in which management wishes to go today is nowhere close to where government legislation is going. Government continues to use words like enforcement, compliance, due diligence, regulations, directives and violations. Management has a fondness for such words as empowerment, responsibility, accountability, employee ownership, quality, customer-centered, quality, service, lean and mean, effectiveness, and rationalize.

These conflicts will continue, and based on the direction in which the world economy appears to be heading, rougher rides are in store as legislators, labor and business attempt to survive and have health and safety play some small role in that survival. Hang on. There will be no easy answers and the ride is going to be long, rough and bumpy, but very interesting.

CHAPTER 7

SAFETY PERFORMANCE MEASUREMENT

Many safety professionals have come to realize that traditional forms of safety measurement have serious deficiencies. Those measures of safety performance which have been used to indicate the levels of safe performance—lost time injuries, frequency and severity, accident costs, etc.—on initial examination, would appear to give a good indication of safety performance. The presumption is that if these numbers are low, they indicate a lack of accidents or injuries, therefore, good safety performance.

The weakness of these conventional after-the-fact accident and injury statistics is that an accident has to first occur in order to have any reliable indicator of performance. This is nothing more than a lesson in bean counting. It tells you nothing about system improvement opportunities or prevention strategies.

Accident and injury indicators (including frequency and severity rates) are failure measures. They have their focus on the negative.

While these result measures are important, they have some severe limitations. They can be used for comparison purposes, but that comparison is still based on failures recorded by your safety system, as compared with your performance, and that of others. And you need to wait until you have a statistically reliable number of these indicators, tracked over a reliable time frame, to give you an accurate picture of your performance. So what should you measure in safety and health systems?

A recent trend has been to utilize safety audits, but these too have limitations. While they are a better form of measurement, and have been deemed to be an indication of a higher standard of care from a due diligence perspective, they have some practical limitations. They are performed infrequently, with various schedules for delivery and cycle time, and they very seldom prescribe an ongoing method of measurement to ensure the issues identified for improvement are being managed effectively.

Usually, another safety audit has to be conducted to determine progress or improvement, or to determine whether the safety audit recommendations have been implemented. Indeed, even the behavioral approach to safety, with its emphasis on at risk or unsafe behaviors or work actions, can be deemed a more pro-active approach to measuring safety, especially if we accept a general definition of safety as being acceptable risk.

MEASURING PROCESS PERFORMANCE

One of the key axioms of business is, “It is very difficult to manage what you can’t measure.” Another popular business phrase still making the rounds today is, “What gets measured gets done.” For these reasons and others, applying the principles of sound performance management is essential for a safety system committed to ongoing improvement.

When we start to critically examine just what it is we should be measuring in our safety systems, we quickly come to a central question: if safety is important, and the issues surrounding the minimization of risk have their logic rooted in the actions of people in the organization (people at all levels), what's important to measure? Additionally, if you're looking for opportunities to offer incentives or recognition options, emphasis needs to be changed from the reactive to the pro-active.

It's important that measures for health, safety and environmental protection be integrated into the strategic plan of each and every business. If there is a risk to the ability of a business to remain competitive and continue to make a profit, measures from a health, safety and environmental perspective need to be used which will not only identify where problems are likely to exist, but will tell how well and effective prevention activities are being managed.

Rather than focus on what has already happened, current and future safety performance measures must focus on improvement of core safety, health and environmental prevention strategies. That's easy to say, but how do you start changing your existing safety culture from one which has depended on accident and injury statistics as the key safety indicators and reward criteria, to one which actively manages prevention efforts on a day-to-day basis and offers achievement-based incentives and recognition accordingly?

The primary objective of performance measurement is to accurately measure how well the organization is accomplishing its objectives, or some other goal or target. Performance measurement can be used to assist in guiding organizational change and performance improvement. When we look at this concept from a safety management perspective, we get a different take on exactly what must be measured and managed in order for safety performance to improve.

There are a number of important steps which you and your organization must first consider before embarking on a strategic safety performance improvement process. These strategic steps include:

1. Before implementing a safety performance improvement system, strategic planning should ideally take place to determine what you want to measure, and why.

What you determine to be important to track and measure should be a direct consequence of the priorities, values, practices and activities which your management team feels will add the most value to the safety improvement effort. You must seriously ask the question, "What do we want to measure, and why?" You also need to ask, "What are we going to do with this information once we are able to capture it?" Ideally, the information you track and measure should be used to allow you to prescribe performance improvement actions, or to reinforce or reward exceptional performance. Once your strategic plan is developed, safety performance measurements can be the basis for assessing whether this strategy is being achieved. Raise the bar on performance expectations and you can offer many safety recognition options and alternatives.

2. Performance Measurement Plan.

The next step in the process is to determine how the measurement system is to be developed and how the system will track ongoing safety management and accident prevention initiatives. What you track and how you will track it will be logical extensions of your strategic safety plan. This step should be undertaken by a senior level management team, and the team should consider the needs of the entire organization in the establishment of the system. This team should consider such factors as measures for success, key performance indicators and their relative importance, and how the information is to be used once it becomes available. This group should be responsible, through consultation and performance management objectives, for laying the foundation for the measurement model and eventual recognition opportunities.

3. Establish Performance Measures, Targets or Standards.

Deciding what you want to measure and why is one of the most important parts of the measurement process. It is a critical stage of the evaluation process that seeks to answer the questions, "What do we want to measure, why, and how important is it to us?" This stage of the process will be critical to the success of your system. At this point, a family of measures should be identified, with a critical evaluation of how these measures fit your entire safety management system. Also critical at this stage is the establishment of specific performance indicators. This involves the development of operational definitions that are specific, agreed upon, and capable of being measured. And the measures are assigned a weight, or value of importance, based on your priorities and values.

Example: Measure the Quality of Your Prevention System

Let's say that one of the key measures identified through your strategic planning is hazard and risk analysis. This element of your safety system has been deemed to be important, and you want to establish a system of ongoing measurement to determine whether it's effective. The first step is to define your standards for risk assessment (when and where they are to be done, by whom, and how often) and then determine exactly what is it that you are going to measure to determine whether the system of risk assessments is working or not. This is not an audit in the traditional sense. This is a system of ongoing appraisal of how well you're managing risk assessment activities.

The following are examples of some of the performance indicators you might use for your risk assessment system:

Safety Performance Measurement: Risk and Hazard Analysis System

- Standards are developed for risk or hazard management—50%
- Jobs are assessed and evaluated for risks or hazards—20%
- Risk or hazard management is used in job training sessions—10%
- Risk or hazard management is used in safety meetings—10%
- Employees assist in the identification of risks or hazards—10%

Once you have determined that these are the measures you will assess, it's then a simple matter of determining what is the best way in which these performance measures are to be tracked, and how will they be valued. One of the more common ways is to assign all measures a total score of 100% and assign each measure a different percentage (or weight) based on the value of these measures to your organization. This will have been accomplished at steps two and three by the strategic planning team.

As you can see from the sample, the total of scores for all the measures associated with the family of risk and hazard analysis is 100%. This value system for the individual measures is based on what a typical organization may define as important, with "standards developed for risk and hazard analysis" being the most important and weighted at 50% to reflect its importance. All other measures in that family are also weighted, based on value or importance.

The key part of the measurement process is tracking on a daily, weekly, monthly or yearly basis how the organization is performing compared to its own standards for risk and hazard analysis. These indicators can be tracked by modifying the forms used by the organization to manage this safety element, and then rating each of the measures and scoring them appropriately.

A key part of the performance measurement system is having a reliable information management system or database to manage this information for ease of data entry and accuracy of reporting. The use of a computer is a necessity to help manage this system.

The number and types of performance measurements can be as many and varied as needed. Use best practice reports and industry research to design your safety management system to be as wide-ranging or as simple as you wish.

4. Establish A Performance Baseline.

Once performance measures are in place, establish a baseline. This will enable you, through the collection of performance measurement data, to measure how effective the system is in achieving its critical success indicators.

5. Use the Measurement Information.

Now simply use the information you have collected to determine your performance. As you can see, the measurement of risk and hazard management initiatives is not dependent on having an accident or injury. You are measuring the effectiveness of the process you have defined for effective risk management, not how many or how few injuries you have.

If you are doing a good job in managing risk, you will have a better indicator of your safety performance. The potential for these accidents and injury statistics to increase or decrease is being managed, and this is what safety performance management and measurement are all about. If you want to offer a safety recognition option, you now have that opportunity based on prevention activities, not mere injury statistics.

SAFETY PERFORMANCE MEASUREMENT OPTIONS FOR STRATEGIC PLANNING AND RECOGNITION

If you really want to turn your safety performance measurement and recognition system into an achievement-based system, rather than an injury-based system, consider some practical measurement and reward alternatives. They take the emphasis off accidents and injuries, and raise the bar on behavior-based safety, inasmuch as they assess not only the behaviors of many people in the business (not simply workers), but the goals, performance targets, standards and achievements of management as well. Additionally, variations on these performance and achievement themes can be part of any systems audit that you decide may be necessary to complement an achievement-based safety model.

They are based on the safety performance measurement software that has been developed by my company, Creative Business Solutions, Inc. The software is called PPM (Performance-Process-Measurement) Safety Management Software, and can ensure maximum flexibility for any company wishing to maximize opportunities to set specific safety performance standards, and provide recognition and feedback on those standards. They lend a whole new dimension to the phrase, "what gets measured gets done." Or preferably, "what gets measured gets managed," because what gets managed stands a better chance of getting recognized, reinforced and repeated.[1]

1. Safety Objective Setting

- Safety objectives in place.
- Safety objectives reviewed periodically.
- Safety objectives are being met.
- Safety objectives shared with employees.
- Safety objectives are both statistical (target) and performance (activity) based.

2. Accident Investigation

- Investigation completed on time.
- Investigation identified the cause(s) of the accident.
- Prevention strategies to prevent recurrence identified.
- Prevention strategies implemented, or in the process of being implemented.

3. Joint Health and Safety Committees

- Committee(s) meet as required.
- Minutes are posted in the workplace as required.

- Equal representation of worker and management representatives.
- Follow up arising from the meeting.
- Follow up completed, or in the process of being completed.

4. Supervisory Development and Safety Management Training

- Safety training needs for management identified.
- Safety training courses for management delivered.
- Senior and middle management have received instruction in legislative compliance and due diligence issues.

5. Inspection and Maintenance

- Inspection/maintenance schedule in place.
- Schedule being followed.
- Inspection/maintenance procedures identify deficiencies or compliance.
- Inspection/maintenance deficiencies being followed up, or compliance recognized.
- Follow up completed, or in the process of being completed.

6. Safety Meetings

- Schedule developed for safety meetings.
- Agenda posted prior to safety meetings.
- Adequate topics prepared/available for safety meeting.
- Current safety performance communicated/updated to staff at safety meeting.
- Follow up from safety meeting completed, or in the process of being completed.

7. Safety Audits

- Safety audits conducted as per audit schedule.
- Deficiencies identified, or compliance recognized.
- Follow up initiated on audit deficiencies.
- Follow up completed on audit deficiencies.
- Compliance recognized.

8. Personal Protective Equipment

- Personal protective equipment needs identified.
- Appropriate personal protective equipment available for job task.
- Appropriate personal protective equipment used as required for job tasks.
- Personal protective equipment maintained appropriately.
- Personal protective equipment stored properly when not in use.

9. Hazard & Risk Analysis

- Jobs assessed and evaluated for risks and hazards.
- Standards developed for risk management.
- Risk management used in job training.
- Risk management used in job planning.
- Employees assist in identification of job risks.

10. Fall Protection

- Falling risks evaluated, using hazard and risk analysis.
- Fall protection standards in place.
- Fall protection equipment available.
- Fall protection equipment being used as required.
- Fall protection equipment properly stored and maintained.

11. Performance Standards—Managerial

- Standards for managerial safety activities defined.
- Standards define frequency of safety activities.
- Standards define responsibilities and accountability.
- Standards define how managerial safety performance is to be measured.
- Performance standards evaluated with each managerial employee.

12. Emergency Response

- Emergency response plans in place.
- Emergency response plans address risks identified in hazard and risk analysis.
- First aid and CPR training needs identified.
- First aid and CPR training conducted as required.

13. Safety Promotion

- Promotional campaigns target specific risk factors.
- Promotional campaigns developed with employee input.

14. Regulatory Compliance System

- Regular assessment of regulatory compliance conducted.
- Regulatory compliance issues discussed at all management meetings.
- Managerial staff get regular updates on regulatory compliance.
- Managerial staff receive instruction on due diligence issues.

15. Pre-Work Planning ("Tool Box" Talks)

- Pre-work plans completed as required.
- Key risk factors identified and minimized through pre-work plan.
- Job completed as per pre-work plan.
- Pre-work plan approved by supervisor on-site.

16. Safe Behavior Observation System

- Safe (at risk) behavior performance standards developed.
- Safe behavior observation conducted as required.
- Safe behaviors noted and recognized for positive reinforcement.
- At risk behaviors noted and corrected.
- Acceptable behaviors documented and reinforced.

17. Contractor Safety Plan and Compliance

- Contractor safety policy in place.
- Contractor safety plan defined for project.
- Safety expectations of contractors defined and shared with contractor.
- Contractor safety performance evaluated.
- Contractor performing project in compliance with contractor safety plan.

18. Housekeeping

- Housekeeping assessment conducted as required.
- Exemplary housekeeping noted and recognized.
- Housekeeping deficiencies identified.
- Corrective action initiated or implemented.

19. Workers' Compensation

- Claims initiated as required.
- Workers' compensation paperwork completed as required.
- Claim managed as per company policies and/or regulatory requirements.

The primary focus here is to develop and implement a strategic safety performance measurement system that will facilitate any type of safety incentive, recognition or award option you deem desirable. The actual type of award may not be as important, in the practical sense, compared to what you actually decide needs to be measured.

There is a difference between making performance improvement measurements in order to determine how effective your business is operating and measuring for reward or recognition purposes. The phenomena known as contest contamination needs to be avoided.

Additionally, if you want to move your safety incentive and recognition system from a failure or accident statistic-based system to an achievement-based model or a performance-based system, the following 100 performance measures, profiled in the September 1995 *Industrial Safety and Hygiene News*, offer opportunity for creativity:

1. Total workers' compensation costs
2. Average cost per claim
3. Costs per man-hour
4. OSHA 200 logs
5. Industry ranking
6. Behavior observation data
7. Bench marking other companies

8. Employee perception surveys
9. Frequency of all injuries/illnesses
10. Severity of all injuries/illnesses
11. Lost time accidents
12. Investigations completed on time
13. Investigation identifies causes
14. Investigation identifies action plan
15. Action plans implemented
16. Safety meetings held as scheduled
17. Agenda promoted in advance
18. Safety records updated and posted
19. Inspections conducted as scheduled
20. Inspection findings brought to closure
21. Management safety communications
22. Management safety participation
23. Near miss/near hit reports
24. Discipline/violation reports
25. Self audits for regulatory compliance
26. Contractor injury/illness statistics
27. Total manufacturing process incidents
28. Total transportation incidents
29. Rate of employee suggestions/complaints
30. Resolution of suggestions/complaints
31. Vehicle accidents per mile driven
32. Safety committee activities
33. Management initiatives
34. Respiratory protection audit
35. Hearing conservation audit
36. Spill control audit
37. Emergency response audit
38. Toxic exposure monitoring audit
39. Ventilation audit
40. Lab safety audit
41. Health/medical services audit
42. Hazard communication audit
43. Ergonomics audit
44. Bloodborne pathogens audit

45. Housekeeping audit
46. Job safety analyses
47. Lockout/tagout audit
48. Confined spaces audit
49. Machine guarding audit
50. Electrical safety audit
51. Vehicle safety audit
52. Fire protection audit
53. Employee participation rates
54. Employee housekeeping
55. Employee safety awareness
56. Employee at risk behavior
57. Supervisor/manager participation
58. Supervisor/manager communication
59. Supervisor/manager enforcement
60. Supervisor/manager safety emphasis
61. Supervisor/manager safety awareness
62. Injury/illness cases reported on time
63. Statistical reports issued on time
64. Ratio of safety and health staff to work force
65. Safety and health spending per employee
66. Titles in safety and health library
67. Technical assistance bulletins issued
68. Policies and procedures updated on time
69. Wellness program participation rates
70. Security audits
71. Emergency drills conducted as planned
72. Percent employees trained in CPR/first aid
73. Absenteeism rates
74. Productivity per employee rates
75. Production error rates
76. Incidence of workplace violence
77. Incidence of accidental releases
78. Employee exit interviews
79. Employee focus groups
80. Community outreach/public safety initiatives
81. Off-the-job safety initiatives

82. Insurance/consultant reports
83. Reports of peer support for safety
84. Certifications of health and safety personnel
85. Percent safety goals achieved
86. Training conducted as scheduled
87. Safety training test scores
88. Statistical tracking of programs
89. Statistical process control
90. System safety analysis
91. Contractor safety activities
92. Positive reinforcement activities
93. OSHA audit—no citations
94. OSHA audit—citations, no fines
95. Willful violations
96. Serious or repeat violations
97. Other-than-serious violations
98. Total dollar amount of penalties
99. Average time to abate reported hazard
100. Average time to respond to complaint [2]

The value of these achievement-based safety criteria is that they can be used to help set performance objectives for everyone in your company, from the most senior executives, to each and every hourly worker. They can help structure your own achievement-based performance model to suit your unique corporate culture, safety goals and objectives.

These measures can be used to complement one another, so that senior and middle management have to support the line in order to achieve their objectives, and workers can see a very definite relationship between their efforts and the corporate direction and philosophy of the business. An achievement-based performance model does not thrive on the exclusion of one group over any other, but is the basis for a high performance system that can energize and strengthen your entire prevention system.

CHAPTER 8

EXPERIENCE WITH INCENTIVE AND RECOGNITION PROGRAMS

Just as there are many and varied opinions on safety incentives, there are also many and varied approaches to their use. Some of these approaches have been implemented with reason and intelligence. Others have been implemented in Skinnerian attempts to modify and control the behavior of others, and some have been implemented with an ill conceived and ill-defined notion of exactly what they were trying to achieve, with predictably poor results.

DO THEY WORK?

To get a flavor of why people either detest or sing the praises of safety incentives, it's useful to go beyond the academic research and so-called controlled experiments of research scientists, and look at what people who have 'been there and done that' have to say about the approach.

In 1991, the Canadian Centre for Occupational Health and Safety published a guideline document for those with an interest in determining if workplace safety incentives were of value. In their document, authors Kevin A. Stewart and Wendy King outlined the case for and against safety incentives. While narrow in its focus due to the fact that the guidelines only looked at employee behavior and incentive strategies, it offered a rudimentary approach to the safety incentive debate. In profiling the case on the pro safety incentive debate, Stewart and King noted the following arguments put forward by supporters of incentives:

1. Safety incentive approaches may help create a positive accident prevention climate.
2. Safety incentives are claimed by supporters to be a method of influencing individual and group work behavior, and making workers more safety conscious.
3. The promotion of safety incentives can generate enthusiasm as part of a strong health and safety program.
4. Some incentive supporters say there is a connection between incentives and having workers act in a desired manner (behavior modification).

Successful incentive programs were deemed to be characterized by the following factors:

1. Good communication and strong efforts aimed at making workers more aware of the actual incentive program.
2. Attainable goals, making it relatively easy for workers to be rewarded for their accomplishments.
3. Non-monetary awards designed to reward as many workers as possible.[1]

The following pro incentive comments reflect the conviction with which supporters embrace the value of incentives:

"Safety research by the Construction Industry Institute through Clemson University shows that the monetary bonus or awards programs are the most effective."

"A yearly incentive bonus would be workable if it includes a reduction factor for the money spent on workers' compensation claims and also certain required measures such as safety meetings and inspections."

"The attitudes of certain employees can be affected by the idea of rewards for being careful at work. Each individual who is motivated to perform work safely makes the entire safety awareness program stronger."

"Non-monetary rewards that apply to all personnel at a location are the best. That approach fosters a team attitude and peer pressure. We have a performance-oriented program based on safety activities, including safety meetings and training, department inspections and safety audits, plus top management participation, and it's working well. A previous program based on days without a lost time accident went sour when people became less than willing to report accidents or felt that other employees held it against them if they did have an accident." [2]

The following opinions and comments on the beneficial aspects of safety incentives were solicited from subscribers to the Internet Safety List (SAFETY@LIST.UVM.EDU):

"Incentives can be useful, and even a little bit of fun, when they are used to draw attention, promote and advertise the overall safety program. They liven up the annual banquet, they provide a tangible symbol of achievement, and so on. But when incentives are the safety program, you don't have much of a program."

"We have a safety incentive program in which working without an accident plays a role. Employees earn safety points for various activities. These are summarized every month and employees can spend their points anytime. We have a catalogue that includes baseball caps, insulated coveralls, tools, binoculars, etc. We believe employees are responsible to work safe, so they should also be responsible to choose their safety incentives. Plus we want to encourage (reward) safe behaviors."

"One of our companies measures by counting the times an employee participates in an audit, an inspection, attends a safety meeting, leads a safety meeting, initiates a safety discovery, etc., and these accumulate and are used in the review process. I realize this may not be measuring the effectiveness of the proactive program, merely the activity taking place, but they have had success in reducing the negative numbers we all seem to have to rely on for our yardstick, and nobody wants to end up with zeros in their safety activity report at review time."

"We still offer incentives for accident free months. We have drawings (for a safety prize) in each department that has accident free months. The prizes are nothing major but the employees enjoy the drawings. National Safety Month is coming in June and we are doing other safety incentives for the month. It works good for us."

"An incentive program puts pressure on workers to work safely, to keep up the team or workforce score, and that's good. On the other hand, it can create a letting-down-the-team feeling when a worker is injured. Not wanting to lower the score, or to hurt teammates, the worker may not report an injury.... When incentive teams are used, the pressure is even greater. If incentives are to be used, let them be individual, or eliminate them altogether."

"What should we be measuring to drive a safety incentive program? Let's assume that safety incentives are valid. We should measure the activities where a change is desired. I would say that safety is desired. So let's measure safe things.... So, I say let's measure the reduction of unsafe activities. Sounds banal up to here. Now, if you must deal with outcomes in some reporting context, do not measure injury outcomes. If you have to measure unwanted outcomes, measure accident and non-accident outcomes. There is a significant difference between an accident and an injury, isn't there?" [3]

... OR DON'T THEY?

Respondents to the 1997 safety incentive survey conducted by *Canadian Occupational Safety* magazine had this to say:

"It's a disincentive to report injuries; it's a gimmick, not a sound health and safety program."

"The union has issued a notice that it will not support programs that would be construed as paying for safety performance."

"People need to accept responsibility for safety without the promise of minor rewards."

"The sad truth is that our department has no incentive for safety, good work or anything else."

"As a government agency, with shrinking budgets, even a modest incentive program carries certain operational costs which at this time we feel could be better served in other areas."

"Health and safety in the workplace is not recognized as moneymaking. All employees can be replaced. In the production manager's eyes, more production is moneymaking and accidents don't matter." [4]

In *The Role of Workplace Safety Incentives*, by Stewart and King, those with negative opinions toward safety incentives claim the following:

1. You cannot buy safety with gifts or others types of incentive approaches. It is much better to have people feel good about their work and their work environment, rather than working towards qualifying for a prize.
2. A program that encourages workers to stay at work when it is unsafe for them to do so because of their physical condition should be avoided. This approach is referred to as the walking wounded, and promotes the underreporting of injuries, not injury prevention.
3. Proper training, effective supervision, knowledge and good human relations motivate more than awards.
4. Reductions in the rate of injuries may be used to show the benefits of safety incentive approaches, but exposures to occupational health and hygiene factors, which may contribute to occupational diseases and illnesses, may not be readily apparent
5. It is difficult to factor into the safety incentive approach differences in job risk, therefore causing indifference, especially among low risk workers.
6. The effect of these types of incentive programs (those which have reduction of injury targets) is to pressure workers to refrain from advising management or committee members about injuries, and having them properly investigated.[5]

The following comments and opinions are from various individuals who have tried the safety incentive approach and found it not workable, unsuitable, or downright useless:

"We had our best year ever for safety in 1996. On a payroll of $2.5 million in a relatively high hazard industry, we had $3,600 in claims costs, give or take a couple of hundred. Having gone on a retro plan in 1996, I had the pleasure of handing my boss a check for $156,000 in workers' compensation premium refunds. I decided to reward my employees by offering them a safety contest in 1997, as recognition for having had such a great year. A lot of planning and thought went into the program, which was adapted from a very successful one in use at a neighboring company (big sign out front, 980 days since the last lost-time accident). It was presented to the executives and other managers at a meeting and approved for a trial period of six months. It involved both team and individual prizes, the employees having voted on their preferred prize, a day off with pay for everyone on the winning team. I put in lots of proactive stuff, such as double points off for unreported incidents and points awarded to the team for safety suggestions and reporting unsafe acts/conditions, attending safety meetings, serving on the safety committee, etc., to get away from the accident-based mentality."

"I was very pleased with the program in its final format. My people proceeded to have accidents in the month of January (average is about .8) and went from there. 1997 was the worst year for claims since 1993 when a guy with a bunion succeeded in getting about $50K in back comp while it was in litigation. We had one serious accident, one really questionable one, and more little ones than usual as well. Not a good year, and I'll be lucky if the refund hits six figures. The program was terminated after the six months, and it will be a cold day in hell before I do another one. The accident rate has, of course, returned to normal. Go figure." [6]

"We can talk all we want about modifying worker behavior and do all we want about reengineering and redesigning the workplace, but until we all learn to investigate incidents and accidents to identify the real root causes, we will be shooting in the dark. How does one prevent these occurrences if we don't look at the real causes? But safety performance with incentives? They look good, sound good and give that warm fuzzy feeling. But they don't work."

"Our company is a safety leader in the utility industry, and we do not have a pay-for-performance safety incentive. The incentive is for our employees to go home alive each day. Safety is the number one priority in this company, and employees believe in the commitment from senior management on down the organization."

"Most employees are safety minded without such incentives. Those who aren't won't change."

"Pay-for-performance plans encourage employees not to report injuries. This underreporting provides inaccurate data and could lead to some serious injuries. Safety should be incorporated into the employee's performance appraisal and should influence the size of the annual raise. This should not include accidents, only a note that the employee had been observed not using safe procedures or methods."

"These plans offer the potential of some minor injuries not being reported and the causes of the injuries not being corrected. The next time an injury occurs, it will be a serious injury or fatality."

"Pay-for-performance plans, especially in a team atmosphere, tend to drive incidents underground. Employees won't report incidents and risk their teammates losing out on a reward. With incidents not being reported, the situation may not be corrected, and a recurrence is bound to happen." [7]

A VETERAN'S PERSPECTIVE

In soliciting feedback and comments on safety incentive programs for this book, one of the more interesting comments came from Bruce Brown, the safety director at Atrium Companies, Inc. of Dallas, Texas. Brown notes [8]:

> "I have been involved with safety incentive programs during my entire safety professional career (about 25 years). During that time, I have accepted programs instituted by my predecessor, purchased programs to fit the need of the company, and have created my own. At one time as a consultant, I was asked to help reduce the number of traffic incidents with the valets at a very well-known hotel, for which I created a very specific incentive program that worked well with them, but would undoubtedly fail if used anywhere else.
>
> Through all this, including creating the safety incentive program for a major defence contractor, I always felt there was a piece missing that made me believe safety incentives were not as effective as vendors of the programs would make you believe. Then at a professional seminar, I heard Bruce Wilkerson state that safety incentives could easily become a 'whose turn is it this month?' program. That was the missing element. As a matter of fact, when I lecture on assessing positives, not negatives, I discuss this and can see the light bulbs flash from people's minds.

Basically, an incentive program is established in a normal department. The department, if indeed normal, consists of varying participants in basically three groups. There are the high achievers, the segment that goes with the flow, and the ones who spend more time making excuses why they can't do something than the time it would take to do it!

The safety incentive program goes into place and immediately you see an improvement. All people, especially the latter category identified, are looking at getting stuff: incentives, money or whatever your program is based on. This may be a long-term improvement on the latter category, bur may not, mainly because the achievers step up slightly and maybe one presents him or herself as the front runner. That person gets the award.

It was expected. All know the achievers. So it is accepted and the others still strive for the next fold ring. Then the same achiever gets the award again the next month or maybe the month after...and all others feel cheated. This may start the loss of interest within the low achievers, as they believe they cannot compete, but may still hang in there. Supervision sees this and understands the wealth must be distributed for the program to work.

The next month, the award goes to one of the middle category. A momentary regeneration follows, but soon dies as the achievers see a prostitution of the award. Everyone knows one of the achievers did more but didn't get the award. The low achievers have some hope but now have settled back into their traditional roles.

Supervision again, in their infinite wisdom, selects one of the lesser achievers the following month and all is lost! The underachievers know it was a gift unearned, and the high achievers see the program as one not based on results and accomplishments, but on 'whose turn is it this month?' The incentive program is now meaningless and ineffective. But the rates are still improving.

Not to be totally negative, but sometimes these programs do function well and survive to be a great catalyst, but they need more than a cursory overview to assess effectiveness. On the positive side, understanding the culture of the workforce and gearing to that from the start can help create a very successful incentive program. By culture, I am not simply talking about the orientation of the workforce but also of the company itself. For instance, if the company gives bonuses and pays well, a monetary incentive program will not have nearly the effect that it would have with a company that is stingy. Many Hispanic cultures are very proud, and an article of clothing that shows their pride in their company will be accepted, whereas many of the Oriental cultures would consider applauding them, with visible symbols in front of their peers, to be out of line.

All-in-all, from my experience, an effective safety incentive program is not just put into place. It is developed from a through knowledge and understanding of what a real incentive is. Getting the working recipients of the program involved from the conception stage will better guarantee acceptance and involvement. Safety incentive can work and work well to lower rates. Just don't put them in place and expect low or not maintenance."

SUCCESS FACTORS

Many people have firm opinions on how to go about structuring a safety incentive program. Fundamental to the establishment and implementation of successful safety incentive, recognition or award programs is knowing the answer to each of the following questions:

1. What exactly is it you hope to achieve with your program?
2. Are you attempting to reduce accident and injury rates, improve simple compliance or conformance to rules, encourage select behaviors, generate management participation, all of these factors or a combination of these factors?
3. How will you, management and workers know when the program is succeeding or failing?
4. Are you sure that one of the many different approaches being marketed by incentive companies is the right approach for you?
5. What are your true feelings about your workers and your management team?
6. What is your current management style, business and regulatory climate? (The better you can accurately describe this factor, the better your incentive, recognition or award strategy will be.)
7. What is your company's perception or opinion of safety management and accident prevention approaches? Do they feel most accidents and injuries are a result of the actions or inactions of workers, or is the entire management system considered when determining accident prevention strategies?

Why would anyone attempt to implement a safety incentive approach for workers, without ever having consulted with them on the need, focus or rationale of the approach? If you think for a minute that this would not happen, you're wrong. While there are more than enough feel good books on the market today about the nirvana of the modern workplace, espousing the virtues of openness, empowerment, enlightenment, and organizational freedom, the reality is that business, including the business of safety management, is hard work. There is far too much literature available which can convince you that a positive safety culture is easy to create and easy to maintain. It's not.

According to Gerald J.S. Wilde, in his book *Target Risk*, while efforts have been made to attempt to cull all the characteristics of an effective incentive plan, the results of these efforts have, to a large degree, been based on inference, due to the fact that there are no well-controlled experiments which favor one particular approach over another. While Wilde may wish to see such data, he laments that it will perhaps never happen, due to the fact that business is not in the habit of running such experiments. It is important to destroy this between the academic, scholarly approach to safety and the real life, ever-changing business realities where the theory about safety incentives sometimes does not match the reality.

The following aspects of planning successful safety incentive programs are based on opinions of academics, social scientists and researchers, consultants, authors and regular, everyday safety professionals:

Successful Safety Incentive Planning

1. Managerial vigor: The program needs to be introduced with consideration for both short and long-term objectives, and should be conducted with managerial vigor, commitment and coherence.
2. Rewarding the bottom line: Incentive programs should reward the outcome and not the process. According to

Wilde, this is because rewarding specific behaviors may not strengthen motivation towards safety, and a potential benefit seen through the reduction of select unsafe behaviors may be offset by a lack of emphasis on other safe behaviors. While some of the rewarded safe behaviors may improve, it may be at the expense of other required safe behaviors.

3. Attractiveness of the award: To take a phrase from the clothing industry, cut the cloth according to the garment. Make sure you have the flexibility to use various types of reinforcement tools, such as praise, trading stamps, lottery tickets, gift certificates, shares of company stock, extra holidays, or merchandise, just to name a few. In other words, you have to figure out the approach which will work for you.
4. Progressive safety credits: The amount of the reward should continue to grow in a progressive fashion, as a greater number of uninterrupted accident-free periods are achieved. Ten years of accident free driving are much harder to achieve than one year of accident free driving.
5. Simple rules: No discussion necessary.
6. Perceived equity: The system has to be fair, and be perceived as being fair. The more you can rationalize the approach, and represent its logic, the better it will stand up to questions, criticism and the day-to-days efforts which people make to contributing to the approach.
7. Perceived attainability: Ensure your reward system is designed in such as way as to convince and promote the idea that the goals and objectives are realistic and attainable.
8. Short incubation period: Make the time frames short so as to maintain interest and keep the focus on the desired outcome. The longer the reward is delayed in being given, the less effective the impact.

9. Reward group as well as individual performance: Keep this in mind if you hope to foster a team approach to safety. Teams are a collection of individuals, and they all have various skills and abilities. The balance of individual and team-based incentives will make for a broader base of appeal, and help build a more solid foundation for a positive safety culture.
10. Participation in the program: Give people a chance to determine what the program should look like, and how it will be administered. This will help foster co-operation. The underlying premise is that people will be more interested in participating in, and working towards, goals and objectives which they've had a hand in developing.
11. Avoidance of accident under-reporting: Make sure your program actually promotes and fosters what you hope it will achieve. Design your system so as to avoid the nonreporting, or underreporting of accidents and injuries.
12. Research the potential impacts of short and long-term feasibility.[9]

According to the information contained in the *Safety Incentive Guide* by the Canadian Centre for Occupational Health and Safety, effective safety incentive approaches are characterized by the following factors:

1. Workers should have an equal opportunity to compete.
2. A committee should be involved in the planning and promotion of the scheme, with members representing all competing groups.
3. The system should be devised so as to ensure that workers have to actively participate in improving safety performance, rather than simply hoping to win something at the end of the day. The objective is to keep the focus on those factors which contribute to enhanced safety, not simply prizes.

4. Group acceptance is important, and approaches to awards should keep worker acceptance in mind. Many opportunities to win should be created, and workers who don't win should not be completely disappointed at the end of the assigned promotion period.
5. Keep it simple and the rules easily understood. If peer pressure becomes an issue, it should be directed in a constructive manner.
6. Themes and attainable goals should be fundamental, with effective use of publicity to create and enhance enthusiasm.
7. Choose awards that are of interest to the award group, make sure you can afford to deliver, and deliver promptly after attainment of the achievement.
8. Carefully plan the length and duration of the promotion to ensure maximum impact, and to avoid having the program go stale and lose interest.[10]

SHOULD YOU HAVE AN INCENTIVE OR RECOGNITION PROGRAM?

In order to make the transition from a failure (statistics) based incentive or recognition culture to an achievement-based safety culture, focus on some very fundamental questions which, when answered, will clearly identify the reasons why a safety incentive or recognition option is even being considered in the first place. These questions include:

1. Why do we want to consider a safety incentive or recognition program for this business?
2. Have we ever tried it in the past? Was it successful? If so, why (and how do we know)? If not, why not (and how do we know)?
3. What is it we hope to achieve with the safety incentive or recognition program?

4. Do we have a problem that we're hoping the safety incentive approach will solve? What is that problem? Are safety incentives and recognition options the correct solution to this problem?
5. Who do we want to be more committed to safety, management or workers?
6. If workers are not participating in our safety system as much as we would like, why not?
7. If management is not participating in our safety system as much as we would like, why not?

Honest answers to these tough questions will bring you closer to deciding if a safety incentive or recognition approach is right for your business. Not only that, you'll have a better idea of exactly which kind of safety incentive approach would be practical, and you'll have the opportunity to decide if a simple statistics approach, an achievement-based approach, or a combination of both is right for you and your business.

CHAPTER 9

EMPLOYER INCENTIVES FOR SAFETY AND COMPLIANCE

Government regulators spend countless hours and resources trying to ensure that employers and workers know, understand and obey the laws and rules put in place to protect the health and safety of millions of employees. To say this is a daunting task is an understatement. To say it is effective can be debated. That there may be better ways is undeniable.

Government's role has been to find employers doing things wrong, not right. Ideally, government should promote safety management excellence, as opposed to simple legislative compliance. Haven't the real success stories in safety come from those businesses which have gone above and beyond the duties of mere compliance and have helped create, shape and sustain a positive safety culture which forces good safety performance?

This point has been raised time and time again by those who have been critical of traditional government approaches to compliance. Just as many businesses are in the process of attempting to change or improve their existing safety culture from one based on negatives (accidents, injuries, breaking rules, non-compliance) to more achievement-based cultures, so too must government keep in touch with the realities of many of the most successful businesses.

MORE BALANCE NEEDED

There should also be recognition and financial incentives for those employees who prove they are top of the list. As I wrote in the July/August 1993 *Canadian Occupational Safety*, based on research conducted by Duke University's W. Kip Viscusi, whether we like it or not, safety sometimes gets traded off, based on the assessment of its worth and value. Viscusi noted that "market forces" have done more for the advancement of safety in the workplace than government agencies like OSHA. According to Viscusi's research, worker's compensation regulations have had roughly ten times the effect on worker safety as OSHA regulations. Viscusi notes, "Ultimately it is the financial incentives created for safety that will affect the tradeoffs that firms make in the promotion of safety. Whether regulatory standards or financial penalty schemes are more effective in creating these incentives depends in large part on the magnitude of the incentives being created. Regulatory standards are only absolute to the extent that they are enforcing absolutely. To date, they have not."[1]

With respect to the economics of safety, Viscusi adds, "Financial incentives do matter, and even from a regulatory standpoint they are a driving force in their promotion of safety. It is for this reason that some economists have suggested that regulatory agencies adopt an injury tax approach to promoting safety rather than relying on the command-and-control regulatory mechanisms." [2]

In a paper presented at the 25th International Congress on Occupational Health in Stockholm, Sweden, in 1966, Ilise Levy-Feitshans and Joseph E. Murphy explored the idea of positive incentives for occupational health, with particular emphasis on workplace violence. Levy-

Feitshans and Murphy noted, "Positive Incentives refers to the use of benefits that motivate corporations or other employer organizations to take effective steps to enhance compliance with law, as a supplement to traditional punishment-and-deterrence models. The latter relies on monetary fines and other penalties after the fact, to punish failure to meet the goals of occupational health and safety compliance. By contrast, positive incentives retain the enforcement features of traditional regulatory models, while also offering an array of programs and rewards that can motivate employers to obey the law and to prevent occupational health and safety hazards even beyond the minimums set by law. With inevitably inadequate resources and a limited array of sanctions, regulators must develop new mechanisms for implementation and compliance. Positive incentives can help meet this demand...."

Levy-Feitshans and Murphy went on to say, "Although it can always be argued that employers need not be rewarded for compliance with the law when they are actually meeting a social responsibility that must be filled anyway regarding compliance, this view understates the magnitude of employer ignorance or non-compliance with occupational safety and health laws and the complexity of factors which lead to employer non-compliance. Recent examples of, or proposals for, positive incentives in the USA's regulatory activity include: public recognition of achievements in the improvement of working conditions; reduced inspections for employers who participate in government-sponsored programs and who maintain levels of compliance specified by those program goals, as in the case of OSHA's Voluntary Protection Program (VPP); streamlined reporting or filing procedures; bidding or purchasing preferences; reduced insurance premiums or assigning leaders in occupational safety and health compliance to special risk pools (this is an especially visible approach for providing incentives for employers to reduce the risk of violence in the workplace); tax credits; and, as OSHA has successfully offered since the 1970's, government funded consultation with immunity from inspection for participants who comply with the consultant's requests, except in cases involving extreme harm. Positive incentives, offering an appropriate mix of rewards and recognition, offer a new approach to prevention that may provide a crucial strategy for curbing workplace violence." [3]

There is no general agreement by the respective players on the value of regulation in effective occupational health and safety performance. Business likes to see fewer restrictions on their practices, claiming that tough regulations make things very hard in a free market economy. Yet regulators and labor state that a world of work without tight, strictly enforced health and safety legislation will result in more injuries and death at work. Many in North America cite Mexico as a prime example of their concern.

As Levy-Feitshans and Murphy have written, "There is international consensus that existing regulatory approaches, relying on punishment as a form of deterrence through fines and other penalties, have been inadequate in preventing occupational safety and health hazards from harming workers. The dilemma is there is nonetheless an absence of political will to impose very tough punishment. Some of the factors that limit the escalation of penalties include:

i) Concern that increased penalties will not increase effectiveness; ii) Concern that negative ramifications within employer organizations will ultimately harm the employability of the workers the laws seek to protect by costing too much money; and iii) Concern that if increased penalties do increase effectiveness of existing regulatory approaches to compliance, then employer organizations would devote greater resources to fighting or gaming the system to resist or circumvent, rather than abide by and support, the laws designed to protect occupational safety and health." [4]

BENEFITS OF POSITIVE INCENTIVES FOR EMPLOYER COMPLIANCE

In the opinion of Levy-Feitshans and Murphy, the opportunity for the concept of positive incentives to play a role in helping employers become more active in the health and safety prevention tasks needs serious consideration.

While many are of the opinion that the only motivation which management will consider is the potential for jail time and fines associated with convictions for safety violations, Levy-Feitshans and Murphy note, "Positive incentives can motivate employers to take steps to prevent accidents in the workplace in several contexts, going beyond compliance with legal minimums. Positive incentives for compliance are open-ended. There are no minimums or maximums. The same incentive can generate new accident and injury prevention programs in the workplace. Positive incentives are sufficiently flexible that they can be combined with other labor-management goals to create a more cooperative atmosphere in the workplace. ... Positive incentives give greater in-house authority to compliance constituencies who may also hold the potential for improved relations between the regulated employer, the workforce, and the government. ... Positive incentives reduce overall enforcement costs, by reducing the use of enforcement mechanisms and by reducing the incidence of violations." [5]

VOLUNTARY PROTECTION OPTIONS

In an effort to recognize and encourage excellence in occupational safety and health protection, the US Department of Labor (OSHA) created the Voluntary Protection Program. Requirements for the system are comprehensive management systems with workers actively involved in anticipating, recognizing, evaluating and controlling potential health and safety hazards at work. Companies wishing to apply for OSHA's VPP must submit a formal application. Employers must identify and describe the key elements of their safety and health management system, including particular emphasis on management leadership and employee involvement, work site analysis, hazard prevention and control, and training. There's even a question or two in the VPP application on positive reinforcement, whereby OSHA asks about the disciplinary system used to enforce safety rules, as well as a general question which asks applicants to "describe any positive reinforcement system you may use." [6]

In the opinion of Levy-Feitshans and Murphy, "Perhaps the most celebrated of all OSHA voluntary programs designed to motivate employers towards effective occupational safety and health protections is the Voluntary Protection Program (VPP) founded in 1982. From its inception, VPP represented the next wave in the trend of experimental enforcement alternatives. Improving upon the pattern of OSHA-funded consultation, VPP represented the evolution into a new generation of interactive programs with a special supportive constituency of its own, specifically the Voluntary Protection Program Participant's Association (VPPPA)."

Under the VPP approach, traditional OSHA consultation

has been enhanced. Also, employer self-inspections have become more formalized. VPP targets employers who already have health and safety management systems in place. It is through the initiative of employers that the process gets supported. OSHA provides assistance through seminars and networking opportunities. Compared to other employers, VPP participants experience lower average lost workday cases.[7]

In a September 26, 1995 speech to the Voluntary Protection Program Participants Association, US Assistant Secretary of Labor for OSHA, Joseph Dear said, "OSHA is known for enforcing safety and health standards: we catch employers doing it wrong." But Dear also went on to describe the new OSHA: an OSHA based on what Dear noted was a three-pronged strategy: offering employers a choice between partnership or traditional enforcement; the use of common sense in developing and enforcing legislation; and a focus on results, not red tape. He also noted it was just as important to "catch employers doing it (safety) right." [8]

But just as the opportunity to explore some alternative enforcement models with variations on the incentive theme seem to be making sense, other forces appear to be at work. On April 7, 1998, US OSHA returned to the traditional enforcement model, foregoing partnerships apparently due to a lack of resources. In a press release headlined, "OSHA Resumes Traditional Enforcement Program; Partnership Remains On Hold", it was noted, "With its premiere new partnership program on hold pending a court ruling, the Occupational Safety and Health Administration (OSHA) today resumes traditional enforcement operations, opening comprehensive inspections in more than 70 cities across the country."

"OSHA Administrator Charles N. Jeffress called the return to traditional enforcement a "second-best but necessary step" for the agency, which had offered voluntary partnerships to 12,000 employers with the highest injury and illness rates in the country...."

"The Cooperative Compliance Program (CCP), unveiled in November 1997, gave employers the option of traditional enforcement or working with OSHA to reduce workplace hazards and receiving a reduced chance of inspection. Of the 12,000 employers invited to join CCP, more than 10,000—or 87 percent—accepted. The partnership program was hailed as a significant step forward for the New OSHA..." [9]

CHAPTER 10

THE FUTURE OF SAFETY AND INCENTIVE STRATEGIES

The coming years are bound to bring many changes in the ways which health and safety strategies are identified, developed, implemented and managed. Significant challenges have arisen as the world of business moves away from its industrial roots to an information and technology driven economy.

Consider your own job for a moment. Is it the same as it was 10 years ago? Five years ago? Six months back? If you're like most, your job and your business have changed dramatically, and will continue to change and evolve.

The information age is creating uncertainty for many workers, including safety professionals. Powerful information management software tools provide are replacing some safety personnel. If health and safety professionals continue to only look at their own narrow areas of expertise, and continue to utilize old tools and techniques which were first popularized 25 years ago, the rest of society will pass us by.

Looking to your own profession for answers to some of the very difficult questions which the current global economy is presenting results in professional in-breeding. As much as it may pain some safety professionals to consider it, the world does not revolve around occupational health and safety. Health and safety issues sometimes are mere reactions to other events and trends unfolding in our businesses and communities. Sometimes those reactions are not as positive as we would like them to be.

In *The End of Work*, Jeremy Rifkin takes a critical look at the changing face of the world of work, describing in frightening detail the decline of the global labor force and the dawn of the post-market era. Stating that the Information Age has arrived, Rifkin chronicles the turbulent upheaval occurring in almost every workplace on the planet.

Despite the potential positive impact of total quality management, teamwork, empowerment and greater employee participation on health and safety in the workplace, there is also a flip side to these issues. Notes Rifkin, "Little has been said or written about the de-skilling of work, the accelerating pace of production, the increased workloads, and the new form of coercion and subtle intimidation that are used to force worker compliance with the requirements of the post-Fordist production practices. ... Of course, it is true that re-engineering and the new information technologies allow companies to collapse layers of management and place more control in the hands of work teams at the point of production. The intent, however, is to increase management's ultimate control over production. Even the effort to solicit the ideas of workers on how to improve performance is designed to increase both the pace and productivity of the plant or office and more fully exploit the full potential of employees." [1]

Not only are we seeing a change in the ways in which work is managed, but also in the very nature of the work itself. As more and more people use computer hardware and software to complete their daily tasks, health and safety is simply no longer about following rules or exhibiting safe behaviors. It is also about dealing with increased stress, less actual control and more perceived control. It is estimated that in the United States alone, work-related stress costs employers in excess of $200 billion a year in absenteeism, reduced productivity, medical expenses and compensation claims.

In the UK, job-stress related losses equal up to 10% of the annual GDP. An International Labor Organization investigator notes that "of all the personal factors related to the causation of accidents, only one emerged as a common denominator, a high level of stress at the time the accident occurred. ... The increased stress levels from working in high-tech, automated work environments is showing up in workers' compensation claims. In 1980 less than five percent of all claims were stress related. By 1989, 15 percent were related to stress disorders." [2]

According to an MIT study, depression costs the American economy $47 billion a year. In Japan, where people work about 4,000 hours a year longer than they do in Europe, they even have a name for death from overwork: *karoshi*.[3]

It's little wonder that some are very cynical about the renewed popularity of the behavior-based approach to safety and behavior modification. If, as many suggest, worker's behavior, attitudes and opinions are simply symptomatic of the business and management environment in which they work, why should the focus be on attempting to change only worker behavior? Indeed, as many studies have clearly pointed out, it is the behavior of many people in the business which, taken together, helps determine whether a positive, effective safety system or culture exists. While a worker behavior-based approach may be appropriate for a rules-based safety management process, if you want to elevate the level of performance to have safety play an important role in overall business strategy, an achievement-based approach is needed.

Think of the potential impact the development of appropriate safety incentive and recognition options which recognize reality, deal with real people, not simply theories of human behavior, and add value to the enterprise, its people, and the bottom line.

X VS. Y

Many recent ideas on how to manage safety have simply come from the popular management books which people, including safety professionals, are reading. What's hot this month? According to Micklethwait and Woolridge, management theorists usually are generally in one of two camps, each driven and guided by a different take on human nature, and each as dramatically different from one another as you can get. Those who follow the doctrine of scientific management believe that the average worker is inherently lazy and greedy. The job of management is to break down jobs into their various parts, so that even the most inept and stupid can master them. And design incentive systems so that even the laziest people will put in some effort. This view has been referred to as Theory X.

Supporters of Theory Y, on the other hand, opt for the humanist approach, purporting that the average worker is intelligent, creative and self-motivating. Management's job is to make sure that the work is both interesting and challenging enough to bring out the best in a worker. This can be accomplished by giving decision making ability to the shop floor worker, creating teams, self-managed or otherwise, and encouraging workers to make suggestions on how the company can improve the way it does business. This has become the debate between hard (scientific) and soft (humanistic) management.[4]

Do businesses, and the people who manage and work for them, have to decide if they are Theory X or Theory Y companies, or can they preach one thing and practice another, a kind of Jekyll and Hyde safety management approach? Too often, businesses are riddled with posters, slogans, clichés, bells and whistles approaches to safety, advocating one thing but practicing another. Shallow attempts at attempting to convince workers to be careful have warped safety cultures from top to bottom. Indeed, many businesses endure roller coaster management, changing and adapting as fads and theories come and go, or the economy changes, or political or social impacts influence the way in which business is conducted.

So how are we to determine just which approaches for creating incentives for enhanced workplace health and safety are correct, and how do we go about recognizing and rewarding those who diligently do an outstanding job? This equation becomes even harder when we look not only at where we've come from, but more importantly where we are going, or would like to be going. While there may be peaks and valleys in respective business cycles, the challenge is to try and minimize the negative side of the cycle, and then maximize the positive, up side.

As Michael Nisbet wrote in the February 16, 1994 edition of the Toronto *Globe and Mail*, "Our pale economy, with its financially undernourished businesses, no longer has enough immunity to resist short-term solutions like Theory X. Back to basics is fashionable. Lean, mean, hard-nosed management is praised; getting tough gets four-star reviews in the financial press. Downsizing, restructuring, reorganizing—all euphemisms for job loss—create an atmosphere of apprehension. When people are frightened, they will endure, at least for a while, getting treated like wheelbarrows. Getting loaded up and pushed around is better than nothing." [5]

Developing and implementing a safety management reinforcement scheme whereby all levels of the organization participate in a constructive way is not as difficult as it may sound. However, one fundamental change must take place in the way in which safety professionals perform their jobs: they must become players in the strategic management process in their respective organizations. Whether they are involved in strategic planning, assisting with setting corporate targets for safety, helping to define performance standards for their management or employee groups, or developing performance measurement and feedback systems, the opportunity to become an integral part of the performance management process, whatever its scope, needs to be created, exploited and utilized.

HIGH PERFORMANCE RECOGNITION AND INCENTIVES

While many of the more traditional safety incentive and award approaches are aimed at getting workers to perform in a specific way (usually select behaviors), it would be foolish for any performance management system to omit or ignore middle, senior and executive management from this safety recognition or incentive mix. That is why the achievement-based safety performance measurement example outlined in Chapter Seven covers all levels of workers, from the senior executives to the hourly worker, and everybody in between.

Following the same logic as the internal responsibility system for occupational health and safety in Canada, the achievement-based safety performance measurement example offers a multi-dimensional approach to strategic safety management, strategic planning, safety improvement targets, goals and objectives, and appropriate rewards and recognition when these targets and objectives are achieved.

The measurement example presents a balanced approach to safety management, and provides opportunities for peer-to-peer recognition, team-based recognition, management driven recognition, and government compliance recognition, all centered on a total safety management approach. No one group is rewarded or recognized to the exclusion of any other. This is a synergistic, prevention and achievement-based approach to safety management.[6]

WHY DOES THE TAIL WAG THE DOG?

Business can show government how a positive, achievement-based safety incentive and recognition system can help improve safety performance. After all, business is dealing with these issues everyday, and success in occupational health and safety, at the end of the day, is the result of workers and management working together on common goals and objectives.

While most of what the trade press has reported about safety incentives more closely resembles folklore than reality, critics are starting to sit up, take notice, and ask some tough questions. As recently as June of 1998, OSHA continued to investigate whether incentive programs discourage workers from reporting injuries. In an interview with *Occupational Hazards* magazine, Rich Fairfax, OSHA's deputy director of compliance noted, "We don't want incentive programs to be an employer's safety and health program because, in many cases, it seems they are actually disincentives to report injuries and illnesses. ... In an absolute worse-case scenario, if there were an incentive program, and injuries and illnesses were not being reported by employees, and the employer didn't know, it could possibly be a General Duty Clause violation." [7]

Organized labor is more blunt. In their opinion (noted in the same article), "Incentive programs ... take worker safety and health out of the hands of employers and place the burden on employees. Labor also believes the programs violate a worker's right to report an injury under the Occupational Safety and Health Act, section 11 c, which deals with whistle blower protection." [8]

COLLABORATION VS. COMPLIANCE

Consider for a moment the comments of Thomas B. Wilson, author of *Innovative Reward Systems for the Changing Workplace*. Wilson notes, "Factors impacting and forcing change in the workplace include an aging workforce and population, changing and more demanding regulatory pressures, a growing concern for the environment, fraud and abuse in financial markets, greater workforce diversity, the speed and scope of information technology and increased competition." [9]

Wilson believes that the incentive, recognition and

reward debate consist of two behavioral viewpoints. The first viewpoint contends that behaviors are best reinforced by work itself and intrinsic motivation drives performance. Alternatively, there are those who believe that rewards and recognition are a necessary condition to achieve or motivate desired behaviors. In a nutshell, those who advocate intrinsic motivation feel that external reinforcement works to reduce creativity and risk taking. Rather than offer variations on traditional carrot and stick approaches, those in the intrinsic motivation camp advise companies to design jobs and working conditions so that people will enjoy the work they do, and the value and enjoyment they receive from the work itself is reward enough. Strong supporters of the recognition and reward approach note their studies and research indicate behavior is a function of its consequences. Positive reinforcement is the most effective way in which to achieve desired behaviors.[10]

CONCLUSION

There, in Wilson's statement, is the entire essence and focus of this book. We've looked at the issues on both sides of the question, and explored challenging, stimulating arguments, both for and against incentive and recognition systems. We, as industrial safety professionals, must raise the bar on safety incentives and recognition. We must consider the benefits of an achievement-based safety culture rather than the traditional injury-based approach, or a singularly focused worker behavior-based incentive or recognition option. And we must challenge business, labor and government to seek new and innovative ways in which to use the power of all our human resources in business.

The next three to five years will see tremendous change

and growth in alternative performance systems, not only within the safety field but within business generally. Alternative forms and methods of compensation will become commonplace. In his article, "Linking Performance Scorecards to Profit-Indexed Performance Pay," William B. Abernathy observed, "While the use of incentive pay has increased among many organizations, a negative relationship can exist between incentive pay and a company's net income when individual performance measures fail to align with organization goals." [11]

The years ahead will see information technology playing a far greater role in safety then ever before. There will even be less for all the staff safety professionals who are employed today to do, but while there will be fewer numbers, those fortunate enough to work in their chosen field of health and safety will have opportunities to contribute to the health and financial integrity of business like never before. The challenge for business, labor and government will be to utilize decades of occupational safety and health experience to move towards a positive, achievement-based safety culture.

APPENDIX A

SAFETY INCENTIVE, RECOGNITION AND AWARD SYSTEM PLANNING

SAFETY INCENTIVES AND RECOGNITION

- Do you presently use them in your business?
- If so, why? If not, why not?
- Reference the definitions of the terms:

  Incentives:

  Rewards:

  Recognition:

- Are they used in any other aspect of your business? If so, how?

ARE SAFETY INCENTIVES AND RECOGNITION RIGHT FOR YOUR BUSINESS?

- Conduct perception surveys—what do people think?

- Consult individual workers & management—What do they *really* think?
- Develop task teams to brainstorm issue.
- Evaluate existing corporate culture of your business.
- Look at existing safety policies and practices: Do they complement a "culture of recognition for outstanding achievement"?
- Are injury rates still the main measure of your safety performance?
- Are there other opportunities to recognize outstanding safety performance?

THE STEPS FOR THE DEVELOPMENT OF A SAFETY INCENTIVE/RECOGNITION SYSTEM

1. Establish the rationale, logic and foundation for the system and its use in your business.

- Will your system be based on injury-free periods only?
- Will the system include select safety and accident prevention activities? If so, what are they?
- Will the system be available to both management and workers?
- Will the system be based on select safety targets, goals and objectives?

- Can workers and management set their own safety objectives?

2. Identify resources at your disposal.

- Will there be a budget for identified rewards or financial remuneration?

- Will the rewards be tied to other forms of compensation, i.e.: executive compensation; pay for performance; merit increases; profit sharing; workers' compensation rebates?

- Are other types of rewards, other than financial, being considered?

- Will resources be available to individuals and teams, and will they be management driven or peer driven?

3. Who is going to drive the process? (Leadership)

- Who do you need to get "on side" to market your cause?

- How will this leadership role be identified?

- Someone with credibility in your business needs to play a key role.

- This leader will need to "sell" others on the positive benefits of an achievement-based safety culture.

- This leader will have to stand firm to the principles and beliefs of an achievement-based safety culture when naysayers looks to find fault with the process.

4. Test marketing: taking the pulse of your people.

- Is the process working?
- Is there a direct cause and effect relationship between your incentive and the safety performance you want/desire?
- Is risk being reduced?
- Does the system need fine-tuning?
- Is the process going stale? If so, what are you doing about it?
- Does the system help foster trust, honest communication, effective safety performance and total safety management objectives? If so, good! If not, why not?

APPENDIX B

SAMPLE SAFETY RECOGNITION AWARD PROGRAM

1.0 INTRODUCTION OF RATIONALE AND PRINCIPLES

The following recommendations are made for the introduction of a safety leadership recognition system for XYZ Company. It's important to remember that safety recognition systems do not replace ongoing accident prevention efforts or due diligence strategies. They simply acknowledge the success of these actions. The following key principles will be paramount for the successful implementation of XYZ Company's safety leadership recognition system:

1.1 Recognition will focus on a combination of region and area team-based achievement opportunities.

1.2 Recognition will focus on those safety factors which XYZ Company management and employees can objectively influence, impact or control. The focus will be on select improvement targets or actions, and shall recognize results. When an achievement is reached, we should be able to identify the efforts or initiatives undertaken, and strategies implemented, which directly contributed to achieving the award. All employees shall have the opportunity to participate in improving many aspects of safety performance.

1.3 The objective shall be to recognize and reward those efforts and activities which are considered important for the company to practice in order to achieve safety excellence. The focus shall be on management leadership, compliance with internal and external safety standards, industry practices and standards, legislative compliance, safe work practices and conditions, feedback and reinforcement.

1.4 Recognition options will be available on a random, monthly and annual basis, thereby providing a broad range of recognition opportunities.

1.5 The focus of safety recognition will include injury reduction targets and accident prevention accomplishments.

2.0 GENERAL GUIDELINES

The following recommendations shall be referenced as general guidelines for the safety leadership recognition effort:

2.1 A region or area shall not be eligible for the President's Award if they experience a fatality in the reporting year.

2.2 It is recommended that the Company Safety Department budget for the President's Award, the Committee of the Year Award and the Quality Investigation Award.

2.3 It is recommended that each region budget for any promotional items they will distribute as recognition items for the Workplace Inspection Excellence Awards.

3.0 SAFETY LEADERSHIP RECOGNITION AWARDS

The following options are recommended.

3.1 Safety Leadership Recognition Award—President's Award

This award will be presented annually to the region that consistently meets outstanding performance targets associated with the Safety Measurement System. A combination of outstanding effort in select accident prevention activities and outstanding frequency, severity and accident cost performance will determine the winner of this award.

To be eligible for this award, each region shall be required to submit a copy of all SMS scores for the year, in addition to a copy of the annual safety action plan for the region, outlining efforts undertaken in the year which contributed to outstanding safety performance, improved knowledge and skills of workers in select safety areas and minimized risk of select hazards. The submission shall also be endorsed by the joint Occupational Health & Safety Committee for the region. The winner of this award will be decided by the president, in consultation with the Safety Department.

The Safety Department recommends that all employees in the winning region receive a high quality, causal style jacket with the safety achievement noted on the garment. However, we would also recommend that if the winning region feels they wish a different tangible award, the region should have the option of selecting an award that they feel would be more appropriate. These awards will be presented at a President's Breakfast, at which time the president will formally acknowledge the achievement and make the presentation of the awards, including the

President's Safety Leadership Plaque for display in the region.

3.2 Safety Leadership Recognition Award—Committee of the Year

Recognizing the contribution which joint health and safety committees can play in an effective safety system, XYZ Company will recognize a Committee of the Year. To be eligible for the Committee of the Year Award, the committee must submit a report of its activities and accomplishments for the year, including an overview of the benefits of committee activities to safety performance in their respective region, including their efforts during North American Occupational Health & Safety Week. The Committee of the Year Award will be determined by representatives from the Safety Department and local labor union. The winner of the XYZ Company Committee of the Year Award will have their award-winning proposal submitted for consideration for any appropriate external awards presented annually to the most pro-active workplace health and safety committee which identifies concerns, implements recommendations, and shows leadership.

3.3 Safety Leadership Recognition Award—Workplace Inspection Excellence

For those crews or individuals which achieve an outstanding compliance rating as part of their inspection under SMS, the supervisor or other individual conducting the inspection shall recognize and promote outstanding safety compliance by distributing a small recognition token (small gift of nominal value) which has a standard, XYZ Company safety message imprinted on the gift. (Gift ideas: baseball hats, pocket knives, flashlights, embroidered safety patch which can be sewn on coveralls, jackets, etc.) These recognition gifts shall be given out, on site, to each crew member, after a perfect inspection compliance rating. The objective is to offer immediate recognition and feedback for safe, risk reducing behaviors and exceptional workplace conditions. An annual Inspection Award of Excellence in each area shall be given to the work crew or individual which achieves a consistently high degree of excellence in the quality of their work practices, compliance and housekeeping.

3.4 Safety Recognition Leadership Award—Quality Investigation Award

While the focus of XYZ Company's safety system is on accident prevention, the reality is that accidents still happen from time to time. As the company now knows, and as SMS now tracks, while accidents will happen and have to be investigated, it's important that when the company does perform investigations that they be of outstanding quality, with their focus on problem solving and improvements. SMS now captures the quality rating of each investigation performed within the company. In addition, a Quality Investigation Recognition Award is proposed.

The Safety Department will evaluate every investigation and will provide feedback to investigation teams that produce superior investigation reports. The awards will be given by the Safety Department and will be based on the criteria of scope and thoroughness of problem solving and analysis, and quality of follow up and prevention initiatives recommended by the investigation team. Each investigation will be evaluated, and on a bi-yearly basis, one investigation will be selected and awarded the Quality Investigation Award.

Upon receipt of the award, the respective investigation team shall be recognized for their quality efforts by the regional manager, with a small recognition gift to mark the achievement, as well as public recognition within their respective region. A photo opportunity and a brief story in the company newsletter, outlining the investigation's benefits to the company shall also be undertaken. Investigations of serious physical injuries or fatalities shall not be considered eligible for this award category.

COMMUNICATION STRATEGY AND INTRODUCTION OF SAFETY LEADERSHIP RECOGNITION SYSTEM

It is recommended that each regional manager and/or superintendent of area operations introduce the concept of the Safety Leadership Recognition Award system, including its rationale, and the local efforts which will be identified or required in order for the respective regions to become eligible to receive the various awards. We would suggest this can be accomplished through regularly scheduled safety meetings, and can start immediately. The Safety Department will put together an overhead presentation and hand-out overview of the approach, and will make sufficient copies of this presentation available to the mangers/superintendents for use during their meetings.

OPTIONAL EXTERNAL SAFETY LEADERSHIP AWARDS

In addition to the safety leadership awards noted above, outside organizations sponsor a number of safety awards. Should the company deem them worthy for consideration, the following types of awards are available which can demonstrate outstanding safety performance for select individual or corporate achievements:

Award of Excellence: Awarded to a company that has demonstrated over time an improved safety record, enhanced product quality, improved training courses and overall company success.

Award of Merit: Awarded to an employee who, through alert and correct action, has saved a person's life or prevented serious workplace injury.

Volunteer Appreciation Award: Awarded to an individual who volunteered his or her time to provide health and safety training to promote an accident free work environment.

Innovative Safety Solution Award: Awarded to individuals or companies that develop innovative solutions for equipment and/or procedures that eliminate safety hazards.

Health and Safety Program Award: Awarded to a company whose leadership and commitment has resulted in the implementation of an outstanding health and safety program.

Should the company wish to apply for any of these awards, applications shall be initiated through the Safety Department.

APPENDIX C

WORKPLACE HEALTH, SAFETY & COMPENSATION COMMISSION OF NEWFOUNDLAND AND LABRADOR LEADERSHIP RECOGNITION AWARDS PROGRAM APPLICATION FORMS

INFORMATION FORM
for the Occupational Health and Safety
LEADERSHIP RECOGNITION AWARDS PROGRAM

Nomination forms must be returned to:
Department of Environment and Labour, P. O. Box 8700, St. John's, NF A1B 4J6
by March 31 of each year

Please complete the section(s) on the following pages for the appropriate award.

If applying for the Minister's Award of Excellence, please complete the following:

Awarded to a company that has demonstrated over time an improved safety record, enhanced product quality, improved training courses and overall company success.

Is there a company policy statement signed by the President? ______

Are there up to date policies and procedure manuals? ______

Are these manuals accessible to all employees? ______

Are the responsibilities for management, supervisors and employees clearly defined? If so, please explain.

______

Does the company have a comprehensive Loss Control Program? If so, please identify the areas that are covered.

______

Has the Loss Control Program been effective in reducing incidents and accidents, and has it created safety awareness within the company? Please explain.

______

Have there been violations under the Occupational Health and Safety Act during the past year with respect to:

Convictions ______
Stop Work Orders ______
Incidence/Accidents ______
Directives Issued ______

Is the company's Safety Program reviewed on an annual basis? If so, please explain. ______

Please list the training courses that are presented to management, supervisors and employees?

______

Has the company pursued the latest advancements in technology to achieve a higher level of productivity? If so, please explain.

______

Has there been an increase in product quality during the past year? If so, please explain.

______

Has the company's workforce expanded to meet the demands of todays economy? If so, explain.

______

If applying for an Award of Merit, please complete the following:

This award will be presented to an employee who, through their actions, has saved a person's life or prevented a serious workplace injury.

Did this incident:

Save a persons life ______

Prevent a serious incident/injury ______

Was the individual given recognition by their employer for their actions? If so, explain.

______

______

Please describe in detail the activities surrounding the incident.

______

______

______

______

If applying for the Volunteer Appreciation Award, please complete the following:

This award will be presented to an individual who volunteered his/her time to provide health and safety training to promote an accident free work environment.

Does the nominee have a consistent record for significant contribution to health and safety? If so, please identify.

______

______

Does the individual volunteer his or her time in the promotion of health and safety? Explain.

______

Does the nominee have professional accreditation? If so, please identify.

______

______

Does the nominee play an active role within an occupational health and safety association? If so, please identify.

______

______

______

In what way has the individual contributed to promoting an accident free workplace? (eg. training, guest speaking, etc.)

______

______

Has the individual shown initiative in designing new training techniques or programs? If so, please identify and indicate if the new procedures have been implemented.

______

______

______

Is the delivery of these programs above and beyond their regular scheduled duties? Explain.

______

______

If applying for Innovative Safety Solution Award, please complete the following:

Awarded to the individuals or companies that develop innovative solutions for equipment and/or procedures that eliminate safety hazards.

Is the innovative safety solution a new system, program or modification to an existing piece of equipment? Please describe.

Has this new safety solution been implemented into the company's safety program?

Are all employees aware and/or receiving training in the proper use of this innovative solution?

Has this new safety concept enhanced the working conditions and increased safety awareness within the company? Please identify the changes that have taken place.

Has there been a reduction in the frequency of injuries/illnesses since the introduction of the new safety idea? Describe.

Please identify the hazards that have been reduced or eliminated since its inception.

Please provide any further information you feel is important to this new safety solution.

If applying for Outstanding Safety Committee/Representative Award, please complete the following:

Does the committee/representative actively identify concerns and implement recommendations to improve the health and safety of employees?

Does the committee maintain records of complaints received from workers concerning health and safety in the workplace?

Are there written procedures which ensure that Occupational Health and Safety committees recommendations are received, considered and properly addressed? Please explain.

Is senior management committed to providing the necessary resources and required training for committee members? If so, how?

Does the committee engage outside consultants such as, new/improved product demonstrators, safety program coordinators or guest speakers to further the knowledge of employees and management on health and safety matters? If so, please provide details.

If applying for Outstanding Safety Committee/Representative Award, please complete the following:

Do the OH&S committees/representatives conduct safety audits of the workplace? If so, identify the areas where audits are performed. If not, why?

Does the committee/representative show leadership in promoting health and safety in the workplace? If so, how?

Is there written procedure in place to provide employees with proper feedback and required information regarding OH&S concerns? If so, is this procedure being followed? Please explain.

Are members of the OH&S committee actively involved in reviewing and revising the companies safety procedures? If so, to what extent?

Does the committee have written procedures for addressing refusal to work situations? If so, are they being followed. Please explain.

If applying for Health and Safety Program Award, please complete the following:

Does the employer have a written safety policy?

Is this policy statement signed by the company president and posted throughout the facility?

Are there written responsibilities for management, supervisors and employees clearly defined?

Is appropriate safety training presented to management, supervisors and employees? Please explain.

Are planned inspections conducted and written reports sent to the Occupational Health and Safety Committee?

Are there written procedures being followed for dealing with hazards reported by employees? If so, explain.

Is there an individual designated as a health and safety coordinator?

Does this individual have the authority to implement corrective measures? If so, explain.

Does the company offer an orientation program for new employees? If so, please identify the areas that this program covers.

Has the implemented safety program had a positive impact on preventing incidence and accidents in the workplace?

Please identify the success of the program and indicate why the program has had such an impact on safety awareness.

Nomination forms must be completed and returned to:
Department of Environment and Labour, P. O. Box 8700, St. John's, NF A1B 4J6
Phone: (709)729-2703 Fax: (709)729 6639
by March 31 of each year

Please place a check in the box beside the award being applied for and complete the following:

- ☐ Award of Merit
- ☐ Achievement Award
- ☐ Innovative Safety Solution Award
- ☐ Outstanding Safety Committee/Representative Award
- ☐ Health and Safety Program Award
- ☐ Minister's Award of Excellence

Name of Nominee ______________________________

Name of Company ______________________________

Type of Industry ______________________________

Has this person been nominated before? __________ If so, when? __________

Address ______________________________

City/Town ______________ Province ______________ Postal Code ______________

Telephone ______________

Name of Nominator ______________________________

Title ______________________________

Company Name ______________________________

Address ______________________________

City/Town ______________ Province ______________ Postal Code ______________

Telephone ______________

Nominator Information
Why do you feel the nominee would be a deserving recipient of a Leadership Recognition Award?

____

____

Please use reverse if additional space is required

______________ Senior Management Representative

______________ OH&S Committee Representative ______________ Chief Executive Officer ______________ OH&S Committee Co Chairs

Please list any attachments you will be providing to either the Nomination Form or the Information Form.

APPENDIX D

OSHA VOLUNTARY PROTECTION PROGRAM INFORMATION AND APPLICATION FORMS

SO YOU WANT TO APPLY TO VPP? HERE'S HOW TO DO IT!

Introduction

We are delighted you are considering applying to the "New OSHA's" Voluntary Protection Programs (VPP). This booklet and its companion piece, *What to Expect During OSHA's Onsite Visit*, have been written to assist you.

The US Department of Labor, OSHA, created the VPP to recognize and encourage excellence in occupational safety and health protection. Requirements for participation are based on comprehensive management systems with employees actively involved in anticipating, recognizing, evaluating, and controlling the potential safety and health hazards at the site.

You may be considering recognition for OSHA's premier program—*Star*—open to companies with comprehensive, successful safety and health programs, who are in the forefront of employee protection. Or you may qualify for *Merit*, designed for firms with a strong commitment to safety and health that need a stepping stone to achieve *Star* performance. Or perhaps you have an alternate approach to safety and health that does not meet all the requirements for the VPP, but still protects your employees at the *Star* level of excellence. Then you would apply to the *Demonstration* VPP.

In any case, applying for VPP is a major undertaking. Preparation of your application will involve a thorough and detailed review of your worksite safety and health program, first by you and then by OSHA. As you conduct your own review, you may find there are gaps that need to be filled before you file a formal application with OSHA.

During the formal application process, OSHA expects each company to describe its program in detail, addressing the required elements for a strong safety and health management system. This booklet outlines the elements OSHA looks for when identifying model companies that fit the VPP profile. In your application, the more clearly you describe how you have implemented these elements at your worksite, the more quickly and appropriately OSHA can respond to your application.

We are glad you are seriously considering VPP, and we look forward to receiving your application. OSHA's VPP is a strong component of the "New OSHA's" commitment to partnership with companies that want to do the right thing—improve worker safety and health.

Instructions

Companies that wish to apply to OSHA's VPP must submit a formal application. This guide is for you to use a workbook in evaluating your current worksite safety and health program. One each page, after a description of the application requirement, there is space for your notes. From these notes, you can develop a comprehensive description of your program that covers all major elements and sub-elements. Most users find that using a team approach to prepare the application works best. If you are taking this approach, you can divide the pages among the team members according to the part of the application each is to complete.

The key elements of an effective safety and health management system—management leadership and employee involvement, worksite analysis, hazard prevention and control, and training—are the focus of this workbook. As you provide the information asked for in this workbook, you will be providing an overview of your company's safety and health activities. This overview will give you a better understanding of your safety and health system's existing strengths and weaknesses and will help you in preparing your application for VPP.

General Information

Site/Company Name

Site Address

Site Manager (Name and Title):

Site VPP Contact (Name and Title):

Phone Numbers:

Company /Corporate Name (if different from above)

Corporate VPP Contact (if applicable—name and title):

Address:

Phone Number:

Collective Bargaining Representative (s)

Union Name and number of Local Chapter:

Name of Union Contact:

Union Address:

Union Phone Number:

Number of Employees

Site employees:

Temporary Employees (routinely on the site):

Contract Employees (routinely on the site):

Type of Work Performed and/or Products Produced

Site's Standard Industrial Code (SIC) - three or four digit number

Injury Rates

- For regular site employees, including temporary employees you supervise, provide the data requested in the following charts for each of the last 3 complete calendar years, plus the average for the 3 years combined.

- For contractors whose employees worked 500 or more hours on your site in any calendar quarter, provide one combined set of data on the charts below reflecting the contractor employee hours worked at your site. (Do not include the hours the contractors employees worked at other sites.)

Injury Incidence Rates

Year	Site Employees: Hours Worked	Site Employees: Recorded Injuries	Rate[1]	Contract Employees: Hours Worked	Contract Employees: Recorded Injuries	Rate
Site 3-year Average[2]						
BLS Industry Average						

Lost-Workday Case Rates[3]

Year	Site Employees: Recorded Lost & Restricted Workday Injuries	Rate[4]	Contract Employees: Recorded Lost & Restricted Workday Injuries	Rate
Site 3-year Average				
BLS Industry Average				

[1]Injury Incidence Rates (IIR) are calculated (N/EH) x 200,000 where:
N = Number of Recordable Injuries in 1 year
EH = Total Number of Hours Worked by all employees in 1 year.
200,000 = Equivalent to 100 full-time workers working 40 hour weeks 50 weeks per year.
[2]Three-year averages are calculated by totalling the hours worked (EH) and the number of injuries (N) and using these total numbers in the formula.
[3]Includes restricted and lost workday injuries.
[4]Calculations are the same as for the IIR except that N = the number of lost and restricted cases.

Management Leadership

Commitment

- Describe your approach to managing the development of your site's safety and health policy.

- Describe your system for communicating this policy to all your employees.

- Describe your system for setting safety and health goals and objectives.

- Describe how these goals and objectives are communicated to all employees.

- Describe how top management is visibly involved in the safety and health program.

- Attach a copy of the site's safety and health policy.
- Attach the current year's safety and health goals and objectives.

Management Leadership

Organization

- Describe how the site safety and health functions fit into your overall management organization.

- Attach an overall organizational chart explaining the relationship of your site's safety and health personnel to your overall organization.

- For large sites, include a separate organizational chart for the safety and health functions.

Responsibility

- Describe the assignment of line and staff safety and health responsibility.

- Attach previously established written material, such as sample job descriptions for managers that include safety and health elements.

Accountability

- Describe the system used for holding line managers and supervisors accountable for safety and health and how that system is documented.

- Attach blank performance appraisal forms for managers and supervisors. (Managers must be evaluated on safety and health performance.)

- Indicate how employees are held accountable for safe and healthful actions.

Management Leadership

Resources

- Provide a narrative summary of personnel, equipment, budget, capital investments (if any), and other resources devoted to your worksite's safety and health management system.

Planning

- Describe how safety and health are a part of your overall management planning, such as setting production goals, increasing or decreasing the workforce, or introducing a new line.

Program Evaluation

All applicants must perform an annual evaluation of the elements required for VPP participation. This evaluation is not the same as a safety audit. It is a review and assessment of the effectiveness of all the program elements: management leadership, employee involvement, worksite analyses, hazard prevention and control, and training

- Describe how you perform the evaluation. For example, who evaluates the program, at what time of year, how is the evaluation report distributed, and how people are held accountable to ensure the recommendations from the evaluation are accomplished?

- Describe how the recommendations from the annual program evaluation are integrated into the safety and health objectives for the next year.

- Attach a copy of the most recent annual evaluation of your entire safety and health program.

- If either the site's or applicable contractor's injury incidence rates and/or the lost-workday case rates are above the national average for the appropriate most recently published SIC, include both long-term and short-term strategies for reducing these rates.*

*Applicable contractors are those who have worked on the site 500 or more hours in any one quarter of the calendar year.

Management Leadership

Contract Workers

- Describe how contractors' past performance in safety and health is taken into account in the bidding process.

- Describe oversight, coordination, and enforcement methods used to ensure that the contractor safety and health program is adequate and is implemented properly. Specify site entry and exit procedures for contractors.

- Describe the means used to ensure prompt correction and/or control of hazards, however detected, under a contractor's control.

- Describe the methods used to ensure that all injuries and illnesses occurring during work performed under a contract are recorded and submitted to you.

- Describe methods, such as monetary penalties and dismissal from the site, used to discourage willful or repeated noncompliance by contractors or their employees.

- List the number of resident contractor companies and the approximate number of contract employees on the site at the time of the application or during the most recent calendar year, whichever most accurately reflects the usual situation at the site.

Employee Notification

- Describe the methods used to ensure that all employees, including newly hired employees, are aware of the following:
  - The site's participation in VPP,
  - An employee's right to file a complaint with OSHA, and
  - An employee's right to receive the results of self-inspections and accident investigations, upon request.

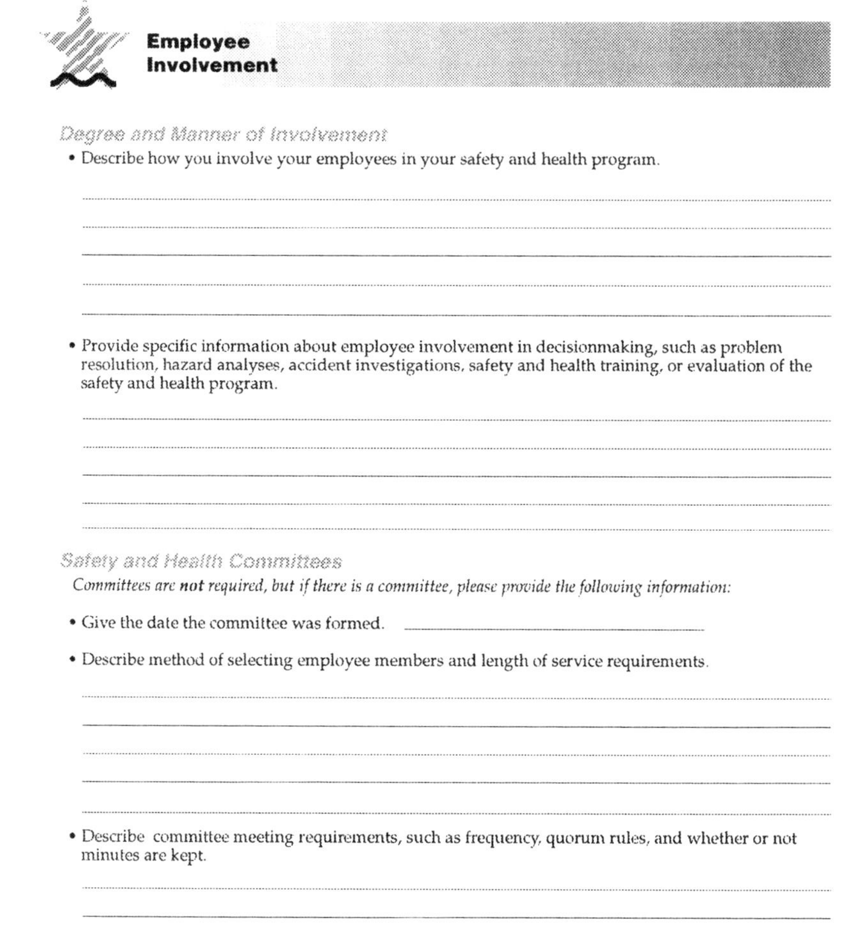

Employee Involvement

Degree and Manner of Involvement

- Describe how you involve your employees in your safety and health program.

- Provide specific information about employee involvement in decisionmaking, such as problem resolution, hazard analyses, accident investigations, safety and health training, or evaluation of the safety and health program.

Safety and Health Committees

*Committees are **not** required, but if there is a committee, please provide the following information:*

- Give the date the committee was formed. ____________________

- Describe method of selecting employee members and length of service requirements.

- Describe committee meeting requirements, such as frequency, quorum rules, and whether or not minutes are kept.

Employee Involvement

- Describe the committee's role in the site's safety and health program, such as:
  - Frequency and scope of committee inspections,
  - Role in accident investigations, and
  - Role in hazard notification.

- Describe hazard recognition training or other specific training for committee members.

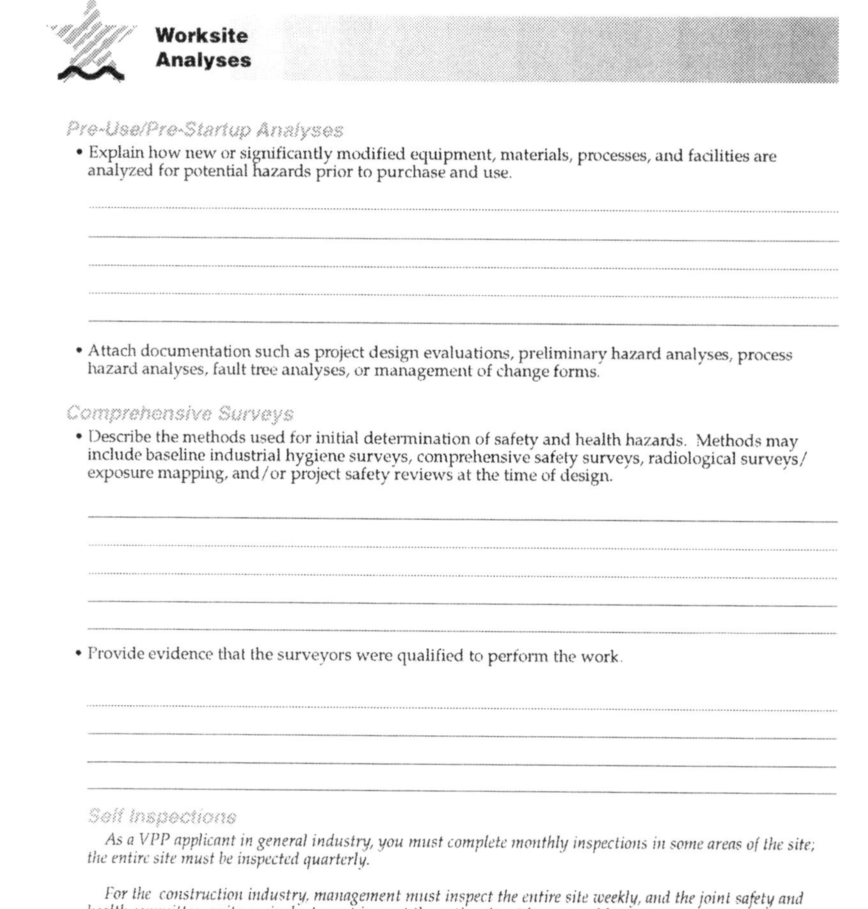

Worksite Analyses

Pre-Use/Pre-Startup Analyses

- Explain how new or significantly modified equipment, materials, processes, and facilities are analyzed for potential hazards prior to purchase and use.

- Attach documentation such as project design evaluations, preliminary hazard analyses, process hazard analyses, fault tree analyses, or management of change forms.

Comprehensive Surveys

- Describe the methods used for initial determination of safety and health hazards. Methods may include baseline industrial hygiene surveys, comprehensive safety surveys, radiological surveys/ exposure mapping, and/or project safety reviews at the time of design.

- Provide evidence that the surveyors were qualified to perform the work.

Self Inspections

As a VPP applicant in general industry, you must complete monthly inspections in some areas of the site; the entire site must be inspected quarterly.

For the construction industry, management must inspect the entire site weekly, and the joint safety and health committee, or its equivalent must inspect the entire site at least monthly.

- Describe the system used to conduct routine, general worksite safety and health inspections. Include schedules and types of inspections, the qualifications of those conducting the inspections, and how corrections are tracked to completion.

- Describe the system for inspecting the entire site quarterly. (For construction, describe the system for inspecting the entire site weekly.)

Routine Hazard Analyses

- State how you review jobs, processes, and/or the interaction among activities to determine safe work procedures at your worksite. Routine procedures, such as job hazard analysis, job safety analysis, and process hazard analysis that result in improved work practices may be used. Describe the frequency of these analyses and provide supporting documentation.

- Describe phase hazard analyses (construction applicants only).

- Describe how results from analyses, such as job hazard analyses, are used in training employees to do their jobs safely and in planning and implementing the hazards correction and control program.

- If process hazards analyses are being conducted, describe how you decide which processes to analyze first.

Worksite Analyses

Employee Reports of Hazards

- Describe how employees notify management when they observe conditions or practices that may pose safety and health hazards. The reporting system must include protection from reprisal, timely and adequate response, and correction of identified hazards.

- Describe how "imminent danger" situations are reported by employees and handled by management.

- Describe how corrections are tracked to completion.

Accident Investigations

- Describe the system used to conduct accident and incident investigations.

Worksite Analyses

- Describe training and/or guidance given to investigators; provide criteria used for deciding which accidents/incidents will be investigated; and describe how near-miss incidents are handled.

- Describe the "lessons learned" process being used at the site and demonstrate root cause analyses.

- Describe the method of tracking recommendations and corrections to completion.

Trend Analyses

- Describe the system(s) used to conduct trend analyses of all data generated by the safety and health program, including employee reports of hazards, hazard assessment data, and injury and illness experience data.

- Describe how the results of the trend analyses are disseminated and used by the line organizations.

Hazard Prevention and Control

Professional Expertise

- Provide details concerning the use of certified professionals, such as occupational medicine personnel, industrial hygienists, and safety professionals in identifying, preventing, and controlling recognized hazards.

- Describe what services are available near or at the site; how these professionals integrate their services with each other; and how communication is maintained.

Safety and Health Rules

- Attach a copy of your site's safety and health rules and describe the disciplinary system used to enforce those rules. Demonstrate that the rules apply to and are communicated to **all** employees.

- Describe any positive reinforcement system you may use.

Personal Protective Equipment

- Describe the requirements for selecting, using, maintaining, and distributing personal protective equipment.

Hazard Prevention and Control

- If respirators are used, attach or list the components or elements of your respiratory protection program—e.g., table of contents of written program. The entire program will be reviewed during the onsite visit.

Preventive Maintenance

- Summarize and briefly describe the procedures you use for preventive maintenance of equipment. Include information on scheduling, and describe how the maintenance timetable is followed.

Emergency Preparedness

- Describe the company's emergency planning and preparedness program. Include information on emergency and annual evacuation drills.

- Describe how credible scenarios are chosen for emergency drills and their relationship to site specific hazards.

Medical Programs

- Describe how you integrate the medical program with the safety and health program.

- Describe the availability of both onsite and offsite medical services or physicians. Indicate the coverage provided by employees trained in first aid, CPR, and other paramedical skills, and indicate what type of training these employees have received. Address coverage for all work shifts.

- Describe how occupational health professionals are involved in routine hazard analyses, in recognizing and treating injuries and illnesses early on, in limiting severity of harm, and in managing injury and illness cases.

Specific Occupational Safety and Health Programs

- List the written occupational safety and health programs implemented at your site, such as a respiratory protection program, where applicable, bloodborne pathogen program, hazard communication program, process safety management program, and lockout/tagout program. (Do not include these programs in your application. They will be reviewed during the onsite visit.)

Training

Employees

- Describe formal and informal safety and health training programs for employees. Specifically address how employees are taught to recognize hazards related to their jobs.

- Describe how often and in what way courses are evaluated and updated.

- Describe testing you use to ensure that employees understand and retain course information.

- Describe how and where training records are kept.

- Describe how frequently training is performed and what prompts repeat training.

Training

Supervisors

- Describe formal and informal safety and health training for supervisors.

Managers

- Describe how top-level managers are trained in their safety and health responsibilities.

OSHA Regional Offices

Region I
(CT,* MA, ME, NH, RI, VT*)
JFK Federal Building
Room E-430
Boston, MA 02203
Telephone: (617) 565-9860

Region II
(NJ, NY,* PR,* VI*)
201 Varick Street
Room 670
New York, NY 10014
Telephone: (212) 337-2378

Region III
(DC, DE, MD,* PA, VA,* WV)
Gateway Building, Suite 2100
3535 Market Street
Philadelphia, PA 19104
Telephone: (215) 596-1201

Region IV
(AL, FL, GA, KY,* MS, NC, SC,* TN*)
Atlanta Federal Center
61 Forsyth Street, SW,
Room 6T50
Atlanta, GA 30303
Telephone: (404) 562-2300

Region V
(IL, IN,* MI,* MN,* OH, WI)
230 South Dearborn Street
Room 3244
Chicago, IL 60604
Telephone: (312) 353-2220

Region VI
(AR, LA, NM,* OK, TX)
525 Griffin Street
Room 602
Dallas, TX 75202
Telephone: (214) 767-4731

Region VII
(IA,* KS, MO, NE)
City Center Square
1100 Main Street, Suite 800
Kansas City, MO 64105
Telephone: (816) 426-5861

Region VIII
(CO, MT, ND, SD, UT,* WY*)
1999 Broadway, Suite 1690
Denver, CO 80202-5716
Telephone: (303) 844-1600

Region IX
(American Samoa, AZ,* CA,* Guam, HI,* NV,* Trust Territories of the Pacific)
71 Stevenson Street
Room 420
San Francisco, CA 94105
Telephone: (415) 975-4310

Region X
(AK,* ID, OR,* WA*)
1111 Third Avenue
Suite 715
Seattle, WA 98101-3212
Telephone: (206) 553-5930

---

*These states and territories operate their own OSHA-approved job safety and health programs (Connecticut and New York plans cover public employees only). States with approved programs must have a standard that is identical to, or at least as effective as, the federal standard.

APPENDIX E

SAFETY INCENTIVE PRODUCTS AND SERVICES

ADMINISTRATION SERVICES

BENNETT BROTHERS, INC., 30 E. Adams St., Chicago, IL 60603, 312-621-1630, FAX: 312-621-1669

BURGER KING CORP., Incentive Sales Dept., 650 E. Devon Ave., Ste.120, Itasca, IL 60143, 800-535-3412

CA SHORT CO., 4205 E. Dixon Blvd., Shelby, NC 28152, 704-482-9591, FAX: 704-484-3749

CERTIF-A-GIFT CO., 4625 25th Ave., Schiller Park, IL 60176, 847-678-3000, 800-323-6849, FAX: 847-678-3806

CREATIVE AWARDS BY LANE, 1575 Elmhurst Rd., Elk Grove, IL 60009, 847-233-5666, FAX: 800-219-8777

CREATIVE OPTIONS USA, INC., 15 Lawrence Bell Dr., Amherst, NY 14221, 800-233-5666, FAX: 800-219-8777

D & L ASSOCIATES, 24795 Country Rd. #75, St. Cloud, MN 56301, 800-328-0307, FAX: 320-252-5504

HAL REED CO., 12004 N.E. 172nd St., Kearney, MO 64060, 816-628-6722, FAX: 816-628-3295

MPC PROMOTIONS, 2026 Shepardsville Rd., Louisville, KY 40218, 502-451-4900, 800-331-0989, FAX: 502-451-5075

MOBIL GO CARD, 650 E. Devon Ave. Ste. 120, Itasca, IL 60143, 800-321-TOGO, FAX: 708-285-6033

NJ & ASSOCIATES, INC., Two Bent Tree Tower, 16479 Dallas Pkwy., Ste. 700, Dallas, TX 75248, 800-622-0177, FAX: 214-248-6626

NATIONAL SAFETY COUNCIL, 1121 Spring Lake Dr., Itasca, IL 60143, 708-285-1121, FAX: 708-285-1315

PROACTION COMMUNICATION SERVICES CORP., 4154 E. Saxony S.E., Grand Rapids, MI 49508, 800-968-6897, FAX: 616-957-5655

SAFETY CONCEPTS, 5241 N. 17th St., Ozark, MO 65721, 417-581-6199, FAX: 417-485-2156

SERVICE MERCHANDISE CO., INC., 7100 Merchandise Dr., Brentwood, TN 37027, 800-367-7375, FAX: 615-660-7632
TOYS "R" US INCENTIVE SALES, 650 E. Devon Ave., Ste. 120, Itasca, IL 60143, 800-887-TOYS, FAX: 708-285-6044

BADGES AND EMBLEM PINS

CREATIVE AWARDS BY LANE, 1575 Elmhurst Rd., Elk Grove, IL 60009, 847-593-1155, FAX: 847-593-1155
CREATIVE OPTIONS USA, INC., 15 Lawrence Bell Dr., Amherst, NY 14221, 800-233-5666, FAX: 800-219-8777
LAB SAFETY SUPPLY, INC., 401 S. Wright Rd., Janesville, WI 53546, 800-356-0783, FAX: 800-543-9910
LABELMASTER, 5724 N. Pulaski Rd., Chicago, IL 60646, 800-621-5808, FAX: 800-723-4327
MPC PROMOTIONS, 2026 Shepardsville Rd., Louisville, KY 40218, 502-451-4900, 800-331-0989, FAX: 502-451-5075
PEAVEY PERFORMANCE SYSTEMS, 14865 W. 105th St., Lenexa, KS 66215, 913-888-1095, FAX: 913-888-3898
POSITIVE IMPRESSIONS INC., 225 Westchester Ave., Port Chester, NY 10573, 914-937-8884, FAX: 914-937-6074
SETON NAME PLATE COMPANY INC., 20 Thompson Rd., Branford, CT 06405, 800-243-6624, FAX: 800-345-7819
WILLIAMS JEWELRY & MFG. CO., 3152 Morris St. N., St. Petersburg, FL 33713, 813-823-7676, FAX: 813-822-2563

BELT BUCKLES

CREATIVE AWARDS BY LANE, 1575 Elmhurst Rd., Elk Grove, IL 60009, 847-233-5666, FAX: 800-219-8777
CREATIVE OPTIONS USA, INC., 15 Lawrence Bell Dr., Amherst, NY 14221, 800-233-5666, FAX: 800-219-8777
D & L ASSOCIATES, 24795 Country Rd. #75, St Cloud, MN 56301, 800-328-0307, FAX: 320-252-5504
MPC PROMOTIONS, 2026 Shepardsville Rd., Louisville, KY 40218, 502-451-4900, 800-331-0989, FAX: 502-451-5075
PEAVEY PERFORMANCE SYSTEMS, 14865 W. 105th St., Lenexa, KS 66215, 913-888-1095, FAX: 913-888-3898
POSITIVE IMPRESSIONS INC., 225 Westchester Ave., Port Chester, NY 10573, 914-937-8884, FAX: 914-937-6074
RAINBOW METALS INC., 17301 Beaton Rd. SE, Monroe, WA 98272, 360-794-3691, FAX: 360-805-0815
WILLIAMS JEWELRY & MFG. CO., 3152 Morris St. N., St. Petersburg, FL 33713, 813-823-7676, FAX: 813-822-2563

CAMERAS

BILL SIMS CO., 102 Lake Vista Dr., Chapin, SC 29063, 803-345-3606, FAX: 803-345-0315
CA SHORT CO., 4205 E. Dixon Blvd., Shelby NC 28152, 704-482-9591, FAX: 704-484-3749
CREATIVE AWARDS BY LANE, 1575 Elmhurst Rd., Elk Grove, IL 60009, 847-593-7700, FAX: 847-593-1155
CREATIVE OPTIONS USA, Inc., 15 Lawrence Bell Dr., Amherst, NY 14221, 800-233-5666, FAX: 800-219-8777
MPC PROMOTIONS, 2026 Shepardsville Rd., Louisville, KY 40218, 502-451-4900, 800-331-0989, FAX: 502-451-5075
PEAVEY PERFORMANCE SYSTEMS, 14865 W. 105th St., Lenexa, KS 66215, 913-888-1095, FAX: 913-888-3898
POSITIVE IMPRESSIONS INC., 225 Westchester Ave., Port Chester, NY 10573, 914-937-8884, FAX: 914-937-6074
SERVICE MERCHANDISE CO., INC., 7100 Merchandise Dr., Brentwood, TN 37027, 800-367-7375, FAX: 615-660-7632
WILLIAMS JEWELRY & MFG. CO., 3152 Morris St. N., St. Petersburg, FL 33713, 813-823-7676, FAX: 813-822-2563

CHINA AND CRYSTAL

CREATIVE AWARDS BY LANE, 1575 Elmhurst Rd., Elk Grove, IL 60009, 847-233-5666, FAX: 800-219-8777
CREATIVE OPTIONS USA, INC., 15 Lawrence Bell Dr., Amherst, NY 14221, 800-233-5666, FAX: 800-219-8777
D & L ASSOCIATES, 24795 Country Rd. #75, St. Cloud, MN 56301, 800-328-0307, FAX: 320-252-5504
HAL REED CO., 12004 N.E. 172nd St., Kearney, MO 64060, 816-628-6722, FAX: 816-628-3295
MPC PROMOTIONS, 2026 Shepardsville Rd., Louisville, KY 40218, 502-451-4900, 800-331-0989, FAX: 502-451-5075
POSITIVE IMPRESSIONS INC., 225 Westchester Ave., Port Chester, NY 10573, 914-937-8884, FAX: 914-937-6074
SERVICE MERCHANDISE CO., INC., 7100 Merchandise Dr., Brentwood, TN 37027, 800-367-7375, FAX: 615-660-7632
WILLIAMS JEWELRY & MFG. CO., 3152 Morris St. N., St. Petersburg, FL 33713, 813-823-7676, FAX: 813-822-2563

CLOCKS

BENNETT BROTHERS, INC., 30 E. Adams St., Chicago, IL 60603, 312-621-1630, FAX: 312-621-1669
BILL SIMS CO., 102 Lake Vista Dr., Chapin, SC 29063, 803-345-3606, FAX: 803-345-0315

BULOVA CORP., One Bulova Ave., Woodside, NY 11377, 718-204-3331, FAX: 718-204-3546
CA SHORT CO., 4205 E. Dixon Blvd., Shelby, NC 28152, 704-482-9591, FAX: 704-484-3749
CREATIVE OPTIONS USA, INC., 15 Lawrence Bell Dr., Amherst, NY 14221, 800-233-5666, FAX: 800-219-8777
D & L ASSOCIATES, 24795 Country Rd. #75, St. Cloud, MN 56301, 800-328-0307, FAX: 320-252-5504
LAB SAFETY SUPPLY, INC., 401 S. Wright Rd., Janesville, WI 53546, 800-356-0783, FAX: 800-543-9910
MPC PROMOTIONS, 2026 Shepardsville Rd., Louisville, KY 40218, 502-451-4900, 800-331-0989, FAX: 502-451-5075
OMNITRAIN/SAFETY SHORTS, 2960 N. 23rd St., La Porte, TX 77571, 800-458-2236, FAX: 713-470-8653
PEAVEY PERFORMANCE SYSTEMS, 14865 W. 105th St., Lenexa, KS 66215, 913-888-1095, FAX: 913-888-3898
POSITIVE IMPRESSIONS INC., 225 Westchester Ave., Port Chester, NY 10573, 914-937-8884, FAX: 914-937-6074
SERVICE MERCHANDISE CO., INC., 7100 Merchandise Dr., Brentwood, TN 37027, 800-367-7375, FAX: 615- 660-7632
WILLIAMS JEWELRY & MFG. CO., 3152 Morris St. N., St. Petersburg, FL 33713, 813-823-7676, FAX: 813-822-2563

CLOTHING (CAPS, JACKETS, SHIRTS, SWEAT-SHIRTS, T-SHIRTS, ETC.)

BILL SIMS CO., 102 Lake Vista Dr., Chapin, SC 29063, 803-345-3606, FAX: 803-345-0315
BRONER GLOVE & SAFETY CO., 1750 Harmon Rd., Auburn Hills, MI 48326, 810-391-5000, FAX: 810-391-5001
CA SHORT CO., 4205 E. Dixon Blvd., Shelby, NC 28152, 704-482-9591, FAX: 704-484-3749
CONNEY SAFETY PRODUCTS, 3202 Latham Dr., Madison, WI 53713, 800-356-9100, FAX: 800-845-9095
CREATIVE AWARDS BY LANE, 1575 Elmhurst Rd., Elk Grove, IL 60009, 847-233-5666, FAX: 800-219-8777
CREATIVE OPTIONS USA, INC., 15 Lawrence Bell Dr., Amherst, NY 14221, 800-233-5666, FAX: 800-219-8777
D & L ASSOCIATES, 24795 Country Rd. #75, St Cloud, MN 56301, 800-328-0307, FAX: 320-252-5504
FIT-RITE HEADWEAR, INC., 92 S. Empire St., Wilkes-Barre, PA 18702, 800-922-5537, FAX: 717-825-8465
HAL REED CO., 12004 N.E. 172nd St., Kearney, MO 64060, 816-628-6722, FAX: 816-628-3295

LL BEAN CORPORATE SALES, Casco St., Freeport, ME 04033, 207-865-4761, 800-832-1889, FAX: 207-797-6585
LAB SAFETY SUPPLY, INC., 401 S. Wright Rd., Janesville, WI 53546, 800-356-0783, FAX: 800-543-9910
LABELMASTER, 5724 N. Pulaski Rd., Chicago, IL 60646, 800-621-5808, FAX: 800-723-4327
ML KISHIGO MFG. CO., 2901 S. Daimler St., Santa Ana, CA 92705, 800-338-9480, FAX: 714-852-0263
MPC PROMOTIONS, 2026 Shepardsville Rd., Louisville, KY 40218, 502-451-4900, 800-331-0989, FAX: 502-451-5075
NATIONAL SAFETY COUNCIL, 1121 Spring Lake Dr., Itasca, IL 60143, 630-285-1121, FAX: 630-285-1315
PEAVEY PERFORMANCE SYSTEMS, 14865 W. 105th St., Lenexa, KS 66215, 913-888-1095, FAX: 913-888-3898
POSITIVE IMPRESSIONS INC., 225 Westchester Ave., Port Chester, NY 10573, 914-937-8884, FAX: 914-937-6074
SAFETY CONCEPTS, 5241 N. 17th St., Ozark, MO 65721, 417-581-6199, FAX: 485-2156
SPIEGEL, INC., 3500 Lacey Rd., Downers Grove, IL 60515, 630-571-8871, 800-982-5664, FAX: 630-571-8860
TJX CORP. INCENTIVES, TJ MAXX/MARSHALLS, 770 Cochituate Rd., PO Box 9360, Framingham, MA 01701, 800-333-1387, FAX: 508-390-5739
WILLIAMS JEWELRY & MFG. CO., 3152 Morris St. N., St. Petersburg, FL 33713, 813-823-7676, FAX: 813-822-2563

COOLERS/ICE CHESTS

CA SHORT CO., 4205 E. Dixon Blvd., Shelby, NC 28152, 704-482-9591, FAX: 704-484-3749
CONNEY SAFETY PRODUCTS, 3202 Latham Dr., Madison, WI 53713, 800-356-9100, FAX: 800-845-9095
CREATIVE AWARDS BY LANE, 1575 Elmhurst Rd., Elk Grove, IL 60009, 847-233-5666, FAX: 800-219-8777
D & L ASSOCIATES, 24795 Country Rd. #75, St. Cloud, MN 56301, 800-328-0307, FAX: 320-252-5504
HAL REED CO., 12004 N.E. 172nd St., Kearney, MO 64060, 816-628-6722, FAX: 816-628-3295
LAB SAFETY SUPPLY, INC., 401 S. Wright Rd., Janesville, WI 53546, 800-356-0783, FAX: 800-543-9910
LABELMASTER, 5724 N. Pulaski Rd., Chicago, IL 60646, 800-621-5808, FAX: 800-723-4327
MPC PROMOTIONS, 2026 Shepardsville Rd., Louisville, KY 40218, 502-451-4900, 800-331-0989, FAX: 502-451-5075

POSITIVE IMPRESSIONS INC., 225 Westchester Ave., Port Chester, NY 10573, 914-937-8884, FAX: 914-937-6074
SERVICE MERCHANDISE CO., INC., 7100 Merchandise Dr., Brentwood, TN 37027, 800-367-7375, FAX: 615- 660-7632
WILLIAMS JEWELRY & MFG. CO., 3152 Morris St. N., St. Petersburg, FL 33713, 813-823-7676, FAX: 813-822-2563

ELECTRONICS (ALARM CLOCKS, ANSWERING MACHINES, PERSONAL STEREOS, RADIOS, STEREOS, TELEPHONES, VIDEOCASSETTE RECORDERS, ETC.)

BEST BUY CO., INC., 7075 Flying Cloud Dr., Eden Prairie, MN 55344, 612-947-2601, FAX: 612-947-2625
CREATIVE AWARDS BY LANE, 1575 Elmhurst Rd., Elk Grove, IL 60009, 847-233-5666, FAX: 800-219-8777
CREATIVE OPTIONS USA, INC., 15 Lawrence Bell Dr., Amherst, NY 14221, 800-233-5666, FAX: 800-219-8777
D & L ASSOCIATES, 24795 Country Rd. #75, St. Cloud, MN 56301, 800-328-0307, FAX: 320-252-5504
HAL REED CO., 12004 N.E. 172nd St., Kearney, MO 64060, 816-628-6722, FAX: 816-628-3295
MPC PROMOTIONS, 2026 Shepardsville Rd., Louisville, KY 40218, 502-451-4900, 800-331-0989, FAX: 502-451-5075
OMNITRAIN/SAFETY SHORTS, 2960 N. 23rd St., La Porte, TX 77571, 800-458-2236, FAX: 713-470-8653
POSITIVE IMPRESSIONS INC., 225 Westchester Ave., Port Chester, NY 10573, 914-937-8884, FAX: 914-937-6074
SAFETY SHORT PRODUCTIONS, 2960 N. 23rd Street, La Porte, TX 77571, 800-458-2236, FAX: 713-470-8653
SERVICE MERCHANDISE CO., INC., 7100 Merchandise Dr., Brentwood, TN 37027, 800-367-7375, FAX: 615- 660-7632
WILLIAMS JEWELRY & MFG. CO., 3152 Morris St. N., St. Petersburg, FL 33713, 813-823-7676, FAX: 813-822-2563

FIRE PROTECTION EQUIPMENT (FIRE BLANKETS, FIRE EXTINGUISHERS, SMOKE DETECTORS, ETC.)

CONNEY SAFETY PRODUCTS, 3202 Latham Dr., Madison, WI 53713, 800-356-9100, FAX: 800-845-9095
CREATIVE AWARDS BY LANE, 1575 Elmhurst Rd., Elk Grove, IL 60009, 847-233-5666, FAX: 800-219-8777
CREATIVE OPTIONS USA, INC., 15 Lawrence Bell Dr., Amherst, NY 14221, 800-233-5666, FAX: 800-219-8777

D & L ASSOCIATES, 24795 Country Rd. #75, St. Cloud, MN 56301, 800-328-0307, FAX: 320-252-5504
WILLIAMS JEWELRY & MFG. CO., 3152 Morris St. N., St. Petersburg, FL 33713, 813-823-7676, FAX: 813-822-2563

FIRST AID KITS

CONNEY SAFETY PRODUCTS, 3202 Latham Dr., Madison, WI 53713, 800-356-9100, FAX: 800-845-9095
CREATIVE AWARDS BY LANE, 1575 Elmhurst Rd., Elk Grove, IL 60009, 847-233-5666, FAX: 800-219-8777
CREATIVE OPTIONS USA, INC., 15 Lawrence Bell Dr., Amherst, NY 14221, 800-233-5666, FAX: 800-219-8777
D & L ASSOCIATES, 24795 Country Rd. #75, St. Cloud, MN 56301, 800-328-0307, FAX: 320-252-5504
HAL REED CO., 12004 N.E. 172nd St., Kearney, MO 64060, 816-628-6722, FAX: 816-628-3295
HEALER PRODUCTS, INC., 3 Rusciano Blvd., Pelham Manor, NY 10803, 800-223-5765, FAX: 914-738-9540
MPC PROMOTIONS, 2026 Shepardsville Rd., Louisville, KY 40218, 502-451-4900, 800-331-0989, FAX: 502-451-5075
POSITIVE IMPRESSIONS INC., 225 Westchester Ave., Port Chester, NY 10573, 914-937-8884, FAX: 914-937-6074
SAF-T-GUARD INTERNATIONAL, INC., 205 Huehl Rd., Dept. B, Northbrook, IL 60062, 847-291-1600, FAX: 847-291-1610
SERVICE MERCHANDISE CO., INC., 7100 Merchandise Dr., Brentwood, TN 37027, 800-367-7375, FAX: 615-660-7632
WILLIAMS JEWELRY & MFG. CO., 3152 Morris St. N., St. Petersburg, FL 33713, 813-823-7676, FAX: 813-822-2563

FLASHLIGHTS

BENNETT BROTHERS, INC., 30 E. Adams St., Chicago, IL 60603, 312-621-1630, FAX: 312-621-1669
CONNEY SAFETY PRODUCTS, 3202 Latham Dr., Madison, WI 53713, 800-356-9100, FAX: 800-845-9095
CREATIVE AWARDS BY LANE, 1575 Elmhurst Rd., Elk Grove, IL 60009, 847-233-5666, FAX: 800-219-8777
CREATIVE OPTIONS USA, INC., 15 Lawrence Bell Dr., Amherst, NY 14221, 800-233-5666, FAX: 800-219-8777
D & L ASSOCIATES, 24795 Country Rd. #75, St. Cloud, MN 56301, 800-328-0307, FAX: 320-252-5504
FULTON INDUSTRIES, INC., 135 E. Linfoot St., Wauseon, OH 43567, 419-335-3015, FAX: 419-335-3215
HAL REED CO., 12004 N.E. 172nd St., Kearney, MO 64060, 816-628-6722, FAX: 816-628-3295

LAB SAFETY SUPPLY, INC., 401 S Wright Rd., Janesville, WI 53546, 800-356-0783, FAX: 800-543-9910
MPC PROMOTIONS, 2026 Shepardsville Rd., Louisville, KY 40218 502-451-4900, 800-331-0989, FAX: 502-451-5075
POSITIVE IMPRESSIONS INC., 225 Westchester Ave., Port Chester, NY 10573, 914-937-8884, FAX: 914-937-6074
SERVICE MERCHANDISE CO., INC., 7100 Merchandise Dr., Brentwood, TN 37027, 800-367-7375, FAX: 615-660-7632
WILLIAMS JEWELRY & MFG. CO., 3152 Morris St. N., St. Petersburg, FL 33713, 813-823-7676, FAX: 813-822-2563

FOODS

BILL SIMS CO., 102 Lake Vista Dr., Chapin, SC 29063, 803-345-3606, FAX: 803-345-0315
BURGER KING CORP., INCENTIVE SALES DEPT., 650 E. Devon Ave., Ste.120, Itasca, IL 60143, 800-535-3412
OMAHA STEAKS, PO Box 3300, Omaha, NE 68103, 800-228-2480, FAX: 800-387-8600
PEAVEY PERFORMANCE SYSTEMS, 14865 W. 105th St., Lenexa, KS 66215, 913-888-1095, FAX: 913-888-3898
STOCK YARDS CHICAGO, 340 N. Oakley Blvd., PO Box 12450, Chicago, IL 60612, 312-733-6050, 800-621-6387, FAX: 312-733-1746
WILLIAMS JEWELRY & MFG. CO., 3152 Morris St. N., St. Petersburg, FL 33713, 813-823-7676, FAX: 813-822-2563

GIFT CERTIFICATES

BENNETT BROTHERS, INC., 30 E. Adams St., Chicago, IL 60603, 312-621-1630, FAX: 312-621-1669
BEST BUY CO., INC., 7075 Flying Cloud Dr., Eden Prairie, MN 55344, 612-947-2601, FAX: 612-947-2625
BILL SIMS CO., 102 Lake Vista Dr., Chapin, SC 29063, 803-345-3606, FAX: 803-345-0315
BURGER KING CORP., Incentive Sales Dept., 650 E. Devon Ave., Ste.120, Itasca, IL 60143, 800-535-3412
CERTIF-A-GIFT CO., 4625 25th Ave., Schiller Park, IL 60176, 847-678-3000, 800-323-6849, FAX: 847-678-3806
CREATIVE AWARDS BY LANE, 1575 Elmhurst Rd., Elk Grove, IL 60009, 847-233-5666, FAX: 800-219-8777
CREATIVE OPTIONS USA, INC., 15 Lawrence Bell Dr., Amherst, NY 14221, 800-233-5666, FAX: 800-219-8777
D & L ASSOCIATES, 24795 Country Rd. #75, St. Cloud, MN 56301, 800-328-0307, FAX: 320-252-5504

IDEA ART, 2603 Elm Hill Pke., Ste. P, Nashville, TN 37214, 615-889-4989, 800-433-2278, FAX: 800-435-2278
JJ KELLER & ASSOCIATES, 3003 W. Breezewood Ln., Neenah, WI 54957, 800-558-5011, FAX: 414-727-7526
LL BEAN CORPORATE SALES, Casco St., Freeport, ME 04033, 207-865-4761, 800-832-1889, FAX: 207-797-6585
LABELMASTER, 5724 N. Pulaski Rd., Chicago, IL 60646, 800-621-5808, FAX: 800-723-4327
MPC PROMOTIONS, 2026 Shepardsville Rd., Louisville, KY 40218, 502-451-4900, 800-331-0989, FAX: 502-451-5075
MOBIL GO CARD, 650 E. Devon Ave. Ste. 120, Itasca, IL 60143, 800-321-TOGO, FAX: 708-285-6033
OMAHA STEAKS, PO Box 3300, Omaha, NE 68103, 800-228-2480, FAX: 800-387-8600
PEAVEY PERFORMANCE SYSTEMS, 14865 W. 105th St., Lenexa, KS 66215, 913-888-1095, FAX: 913-888-3898
POSITIVE IMPRESSIONS INC., 225 Westchester Ave., Port Chester, NY 10573, 914-937-8884, FAX: 914-937-6074
PREFERRED PROMOTIONS NETWORKS, PO Box 1830, Stephenville, TX 76401, 254-965-7792, 800-752-1210, FAX: 254-965-7793
SERVICE MERCHANDISE CO., INC., 7100 Merchandise Dr., Brentwood, TN 37027, 800-367-7375, FAX: 615-660-7632
SPIEGEL, INC., 3500 Lacey Rd., Downers Grove, IL 60515, 630-571-8871, 800-982-5664, FAX: 630-571-8860
STOCK YARDS CHICAGO, 340 N. Oakley Blvd., PO Box 12450, Chicago, IL 60612, 312-733-6050, 800-621-6387, FAX: 312-733-1746
STORED VALUE MARKETING, 4825 N. Scott St., Ste. 100, Schiller Park, IL 60176, 847-671-1300, 800-972-7481, FAX: 847-671-5060
TJX CORP. INCENTIVES, TJ MAXX/MARSHALLS, 770 Cochituate Rd., PO Box 9360, Framingham, MA 01701, 800-333-1387, FAX: 508-390-5739
TOYS "R" US INCENTIVE SALES, 650 E. Devon Ave., Ste. 120, Itasca, IL 60143, 800-887-TOYS, FAX: 708-285-6044

HIGHWAY EMERGENCY KITS

CONNEY SAFETY PRODUCTS, 3202 Latham Dr., Madison, WI 53713, 800-356-9100, FAX: 800-845-9095
HAL REED CO., 12004 N.E. 172nd St., Kearney, MO 64060, 816-628-6722, FAX: 816-628-3295
NATIONAL MARKER CO., PO Box 1659, Pawtucket, RI 02862, 800-453-2727, FAX: 800-338-0309

POSITIVE IMPRESSIONS INC., 225 Westchester Ave., Port Chester, NY 10573, 914-937-8884, FAX: 914-937-6074
WILLIAMS JEWELRY & MFG. CO., 3152 Morris St. N., St. Petersburg, FL 33713, 813-823-7676, FAX: 813-822-2563

INCENTIVE PACKAGES (PRICE RANGES: $10 TO OVER $1,000)

BENNETT BROTHERS, INC., 30 E. Adams St., Chicago, IL 60603, 312-621-1630, FAX: 312-621-1669
BULOVA CORP., One Bulova Ave., Woodside, NY 11377, 718-204-3331, FAX: 718-204-3546
CA SHORT CO., 4205 E. Dixon Blvd., Shelby, NC 28152, 704-482-9591, FAX: 704-484-3749
CERTIF-A-GIFT CO., 4625 25th Ave., Schiller Park, IL 60176, 847-678-3000, 800-323-6849, FAX: 847-678-3806
CREATIVE AWARDS BY LANE, 1575 Elmhurst Rd., Elk Grove, IL 60009, 847-233-5666, FAX: 800-219-8777
CREATIVE OPTIONS USA, INC., 15 Lawrence Bell Dr., Amherst, NY 14221, 800-233-5666, FAX: 800-219-8777
D & L ASSOCIATES, 24795 Country Rd. #75, St. Cloud, MN 56301, 800-328-0307, FAX: 320-252-5504
HAL REED CO., 12004 N.E. 172nd St., Kearney, MO 64060, 816-628-6722, FAX: 816-628-3295
MPC PROMOTIONS, 2026 Shepardsville Rd., Louisville, KY 40218, 502-451-4900, 800-331-0989, FAX: 502-451-5075
PEAVEY PERFORMANCE SYSTEMS, 14865 W. 105th St., Lenexa, KS 66215, 913-888-1095, FAX: 913-888-3898
POSITIVE IMPRESSIONS INC., 225 Westchester Ave., Port Chester, NY 10573, 914-937-8884, FAX: 914-937-6074
SAFETY CONCEPTS, 5241 N. 17th St., Ozark, MO 65721, 417-581-6199, FAX: 485-2156
SERVICE MERCHANDISE CO., INC., 7100 Merchandise Dr., Brentwood, TN 37027, 800-367-7375, FAX: 615-660-7632

JEWELRY (BRACELETS, NECKLACES, PINS, RINGS, TIE TACKS, ETC.)

BENNETT BROTHERS, INC., 30 E. Adams St., Chicago, IL 60603, 312-621-1630, FAX: 312-621-1669
BILL SIMS CO., 102 Lake Vista Dr., Chapin, SC 29063, 803-345-3606, FAX: 803-345-0315
CREATIVE AWARDS BY LANE, 1575 Elmhurst Rd., Elk Grove, IL 60009, 847-233-5666, FAX: 800-219-8777

CREATIVE OPTIONS USA, INC., 15 Lawrence Bell Dr., Amherst, NY 14221, 800-233-5666, FAX: 800-219-8777
D & L ASSOCIATES, 24795 Country Rd. #75, St. Cloud, MN 56301, 800-328-0307, FAX: 320-252-5504
OMNITRAIN/SAFETY SHORTS, 2960 N. 23rd St., La Porte, TX 77571, 800-458-2236, FAX: 713-470-8653
PEAVEY PERFORMANCE SYSTEMS, 14865 W. 105th St., Lenexa, KS 66215, 913-888-1095, FAX: 913-888-3898
POSITIVE IMPRESSIONS INC., 225 Westchester Ave., Port Chester, NY 10573, 914-937-8884, FAX: 914-937-6074
SERVICE MERCHANDISE CO., INC., 7100 Merchandise Dr., Brentwood, TN 37027, 800-367-7375, FAX: 615-660-7632
WILLIAMS JEWELRY & MFG. CO., 3152 Morris St. N., St. Petersburg, FL 33713, 813-823-7676, FAX: 813-822-2563

KEY CHAINS

CREATIVE AWARDS BY LANE, 1575 Elmhurst Rd., Elk Grove, IL 60009, 847-233-5666, FAX: 800-219-8777
CREATIVE OPTIONS USA, INC., 15 Lawrence Bell Dr., Amherst, NY 14221, 800-233-5666, FAX: 800-219-8777
D & L ASSOCIATES, 24795 Country Rd. #75, St. Cloud, MN 56301, 800-328-0307, FAX: 320-252-5504
HAL REED CO., 12004 N.E. 172nd St., Kearney, MO 64060, 816-628-6722, FAX: 816-628-3295
LAB SAFETY SUPPLY, INC., 401 S. Wright Rd., Janesville, WI 53546, 800-356-0783, FAX: 800-543-9910
MPC PROMOTIONS, 2026 Shepardsville Rd., Louisville, KY 40218, 502-451-4900, 800-331-0989, FAX: 502-451-5075
NATIONAL SAFETY COUNCIL, 1121 Spring Lake Dr., Itasca, IL 60143, 630-285-1121, FAX: 630-285-1315
POSITIVE IMPRESSIONS INC., 225 Westchester Ave., Port Chester, NY 10573, 914-937-8884, FAX: 914-937-6074
WILLIAMS JEWELRY & MFG. CO., 3152 Morris St. N., St. Petersburg, FL 33713, 813-823-7676, FAX: 813-822-2563

LUGGAGE

BILL SIMS CO., 102 Lake Vista Dr., Chapin, SC 29063, 803-345-3606, FAX: 803-345-0315
CREATIVE AWARDS BY LANE, 1575 Elmhurst Rd., Elk Grove, IL 60009, 847-233-5666, FAX: 800-219-8777
CREATIVE OPTIONS USA, INC., 15 Lawrence Bell Dr., Amherst, NY 14221, 800-233-5666, FAX: 800-219-8777
D & L ASSOCIATES, 24795 Country Rd. #75, St. Cloud, MN 56301, 800-328-0307, FAX: 320-252-5504

HAL REED CO., 12004 N.E. 172nd St., Kearney, MO 64060, 816-628-6722, FAX: 816-628-3295
MPC PROMOTIONS, 2026 Shepardsville Rd., Louisville, KY 40218, 502-451-4900, 800-331-0989, FAX: 502-451-5075
PEAVEY PERFORMANCE SYSTEMS, 14865 W. 105th St., Lenexa, KS 66215, 913-888-1095, FAX: 913-888-3898
POSITIVE IMPRESSIONS INC., 225 Westchester Ave., Port Chester, NY 10573, 914-937-8884, FAX: 914-937-6074
SERVICE MERCHANDISE CO., INC., 7100 Merchandise Dr., Brentwood, TN 37027, 800-367-7375, FAX: 615- 660-7632
WILLIAMS JEWELRY & MFG. CO., 3152 Morris St. N., St. Petersburg, FL 33713, 813-823-7676, FAX: 813-822-2563

MUSIC (CD'S, CASSETTES, TAPES)

SONY MUSIC SPECIAL PRODUCTS, 550 Madison Ave., New York, NY 10022, 212-833-8655, 800-356-1052, FAX: 212-833-7021

PATCHES (EMBROIDERED)

CREATIVE AWARDS BY LANE, 1575 Elmhurst Rd, Elk Grove, IL 60009, 847-233-5666, FAX: 800-219-8777
CREATIVE OPTIONS USA, INC., 15 Lawrence Bell Dr., Amherst, NY 14221, 800-233-5666, FAX: 800-219-8777
MPC PROMOTIONS, 2026 Shepardsville Rd., Louisville, KY 40218, 502-451-4900, 800-331-0989, FAX: 502-451-5075
POSITIVE IMPRESSIONS INC., 225 Westchester Ave., Port Chester, NY 10573, 914-937-8884, FAX: 914-937-6074
WILLIAMS JEWELRY & MFG. CO., 3152 Morris St. N., St. Petersburg, FL 33713, 813-823-7676, FAX: 813-822-2563

PENS

CREATIVE AWARDS BY LANE, 1575 Elmhurst Rd., Elk Grove, IL 60009, 847-233-5666, FAX: 800-219-8777
CREATIVE OPTIONS USA, INC., 15 Lawrence Bell Dr., Amherst, NY 14221, 800-233-5666, FAX: 800-219-8777
D & L ASSOCIATES, 24795 Country Rd. #75, St. Cloud, MN 56301, 800-328-0307, FAX: 320-252-5504
HAL REED CO., 12004 N.E. 172nd St., Kearney, MO 64060, 816-628-6722, FAX: 816-628-3295
LAB SAFETY SUPPLY, INC., 401 S. Wright Rd., Janesville, WI 53546, 800-356-0783, FAX: 800-543-9910
LABELMASTER, 5724 N. Pulaski Rd., Chicago, IL 60646, 800-621-5808, FAX: 800-723-4327
MPC PROMOTIONS, 2026 Shepardsville Rd., Louisville, KY 40218, 502-451-4900, 800-331-0989, FAX: 502-451-5075

CREATIVE OPTIONS USA, INC., 15 Lawrence Bell Dr., Amherst, NY 14221, 800-233-5666, FAX: 800-219-8777
D & L ASSOCIATES, 24795 Country Rd. #75, St. Cloud, MN 56301, 800-328-0307, FAX: 320-252-5504
OMNITRAIN/SAFETY SHORTS, 2960 N. 23rd St., La Porte, TX 77571, 800-458-2236, FAX: 713-470-8653
PEAVEY PERFORMANCE SYSTEMS, 14865 W. 105th St., Lenexa, KS 66215, 913-888-1095, FAX: 913-888-3898
POSITIVE IMPRESSIONS INC., 225 Westchester Ave., Port Chester, NY 10573, 914-937-8884, FAX: 914-937-6074
SERVICE MERCHANDISE CO., INC., 7100 Merchandise Dr., Brentwood, TN 37027, 800-367-7375, FAX: 615-660-7632
WILLIAMS JEWELRY & MFG. CO., 3152 Morris St. N., St. Petersburg, FL 33713, 813-823-7676, FAX: 813-822-2563

KEY CHAINS

CREATIVE AWARDS BY LANE, 1575 Elmhurst Rd., Elk Grove, IL 60009, 847-233-5666, FAX: 800-219-8777
CREATIVE OPTIONS USA, INC., 15 Lawrence Bell Dr., Amherst, NY 14221, 800-233-5666, FAX: 800-219-8777
D & L ASSOCIATES, 24795 Country Rd. #75, St. Cloud, MN 56301, 800-328-0307, FAX: 320-252-5504
HAL REED CO., 12004 N.E. 172nd St., Kearney, MO 64060, 816-628-6722, FAX: 816-628-3295
LAB SAFETY SUPPLY, INC., 401 S. Wright Rd., Janesville, WI 53546, 800-356-0783, FAX: 800-543-9910
MPC PROMOTIONS, 2026 Shepardsville Rd., Louisville, KY 40218, 502-451-4900, 800-331-0989, FAX: 502-451-5075
NATIONAL SAFETY COUNCIL, 1121 Spring Lake Dr., Itasca, IL 60143, 630-285-1121, FAX: 630-285-1315
POSITIVE IMPRESSIONS INC., 225 Westchester Ave., Port Chester, NY 10573, 914-937-8884, FAX: 914-937-6074
WILLIAMS JEWELRY & MFG. CO., 3152 Morris St. N., St. Petersburg, FL 33713, 813-823-7676, FAX: 813-822-2563

LUGGAGE

BILL SIMS CO., 102 Lake Vista Dr., Chapin, SC 29063, 803-345-3606, FAX: 803-345-0315
CREATIVE AWARDS BY LANE, 1575 Elmhurst Rd., Elk Grove, IL 60009, 847-233-5666, FAX: 800-219-8777
CREATIVE OPTIONS USA, INC., 15 Lawrence Bell Dr., Amherst, NY 14221, 800-233-5666, FAX: 800-219-8777
D & L ASSOCIATES, 24795 Country Rd. #75, St. Cloud, MN 56301, 800-328-0307, FAX: 320-252-5504

HAL REED CO., 12004 N.E. 172nd St., Kearney, MO 64060, 816-628-6722, FAX: 816-628-3295
MPC PROMOTIONS, 2026 Shepardsville Rd., Louisville, KY 40218, 502-451-4900, 800-331-0989, FAX: 502-451-5075
PEAVEY PERFORMANCE SYSTEMS, 14865 W. 105th St., Lenexa, KS 66215, 913-888-1095, FAX: 913-888-3898
POSITIVE IMPRESSIONS INC., 225 Westchester Ave., Port Chester, NY 10573, 914-937-8884, FAX: 914-937-6074
SERVICE MERCHANDISE CO., INC., 7100 Merchandise Dr., Brentwood, TN 37027, 800-367-7375, FAX: 615- 660-7632
WILLIAMS JEWELRY & MFG. CO., 3152 Morris St. N., St. Petersburg, FL 33713, 813-823-7676, FAX: 813-822-2563

MUSIC (CD'S, CASSETTES, TAPES)

SONY MUSIC SPECIAL PRODUCTS, 550 Madison Ave., New York, NY 10022, 212-833-8655, 800-356-1052, FAX: 212-833-7021

PATCHES (EMBROIDERED)

CREATIVE AWARDS BY LANE, 1575 Elmhurst Rd, Elk Grove, IL 60009, 847-233-5666, FAX: 800-219-8777
CREATIVE OPTIONS USA, INC., 15 Lawrence Bell Dr., Amherst, NY 14221, 800-233-5666, FAX: 800-219-8777
MPC PROMOTIONS, 2026 Shepardsville Rd., Louisville, KY 40218, 502-451-4900, 800-331-0989, FAX: 502-451-5075
POSITIVE IMPRESSIONS INC., 225 Westchester Ave., Port Chester, NY 10573, 914-937-8884, FAX: 914-937-6074
WILLIAMS JEWELRY & MFG. CO., 3152 Morris St. N., St. Petersburg, FL 33713, 813-823-7676, FAX: 813-822-2563

PENS

CREATIVE AWARDS BY LANE, 1575 Elmhurst Rd., Elk Grove, IL 60009, 847-233-5666, FAX: 800-219-8777
CREATIVE OPTIONS USA, INC., 15 Lawrence Bell Dr., Amherst, NY 14221, 800-233-5666, FAX: 800-219-8777
D & L ASSOCIATES, 24795 Country Rd. #75, St. Cloud, MN 56301, 800-328-0307, FAX: 320-252-5504
HAL REED CO., 12004 N.E. 172nd St., Kearney, MO 64060, 816-628-6722, FAX: 816-628-3295
LAB SAFETY SUPPLY, INC., 401 S. Wright Rd., Janesville, WI 53546, 800-356-0783, FAX: 800-543-9910
LABELMASTER, 5724 N. Pulaski Rd., Chicago, IL 60646, 800-621-5808, FAX: 800-723-4327
MPC PROMOTIONS, 2026 Shepardsville Rd., Louisville, KY 40218, 502-451-4900, 800-331-0989, FAX: 502-451-5075

NATIONAL SAFETY COUNCIL, 1121 Spring Lake Dr., Itasca, IL 60143, 630-285-1121, FAX: 630-285-1315
PEAVEY PERFORMANCE SYSTEMS, 14865 W. 105th St., Lenexa, KS 66215, 913-888-1095, FAX: 913-888-3898
POSITIVE IMPRESSIONS INC., 225 Westchester Ave., Port Chester, NY 10573, 914-937-8884, FAX: 914-937-6074
SERVICE MERCHANDISE CO., INC., 7100 Merchandise Dr, Brentwood, TN 37027, 800-367-7375, FAX: 615- 660-7632
WILLIAMS JEWELRY & MFG. CO., 3152 Morris St. N., St. Petersburg, FL 33713, 813-823-7676, FAX: 813-822-2563

PINS

CREATIVE AWARDS BY LANE, 1575 Elmhurst Rd., Elk Grove, IL 60009, 847-233-5666, FAX: 800-219-8777
CREATIVE OPTIONS USA, INC., 15 Lawrence Bell Dr., Amherst, NY 14221, 800-233-5666, FAX: 800-219-8777
D & L ASSOCIATES, 24795 Country Rd. #75, St. Cloud, MN 56301, 800-328-0307, FAX: 320-252-5504
LAB SAFETY SUPPLY, INC., 401 S. Wright Rd., Janesville, WI 53546, 800-356-0783, FAX: 800-543-9910
LABELMASTER, 5724 N. Pulaski Rd., Chicago, IL 60646, 800-621-5808, FAX: 800-723-4327
MPC PROMOTIONS, 2026 Shepardsville Rd., Louisville, KY 40218, 502-451-4900, 800-331-0989, FAX: 502-451-5075
NATIONAL SAFETY COUNCIL, 1121 Spring Lake Dr., Itasca, IL 60143, 630-285-1121, FAX: 630-285-1315
POSITIVE IMPRESSIONS INC., 225 Westchester Ave., Port Chester, NY 10573, 914-937-8884, FAX: 914-937-6074
RAINBOW METALS INC., 17301 Beaton Rd. S.E., Monroe, WA 98272, 360-794-3691, FAX: 360-805-0815
SETON NAME PLATE CO., 20 Thompson Rd., Branford, CT 06405, 800-243-6624, FAX: 800-345-7819
WILLIAMS JEWELRY & MFG. CO., 3152 Morris St. N., St. Petersburg, FL 33713, 813-823-7676, FAX: 813-822-2563

PLANT SAFETY SCOREBOARD (NUMBER OF DAYS WORKED WITHOUT A LOST-TIME INJURY)

GRANDWELL INDUSTRIES, INC., 121 Quantum St., Holly Springs, NC 27540, 919-557-1221, FAX: 919-552-9830
LAB SAFETY SUPPLY, INC., 401 S. Wright Rd., Janesville, WI 53546, 800-356-0783, FAX: 800-543-9910
MPC PROMOTIONS, 2026 Shepardsville Rd., Louisville, KY 40218, 502-451-4900, 800-331-0989, FAX: 502-451-5075

NATIONAL MARKER COMPANY, P.O. Box 1659, Pawtucket, RI 02862, 800-338-0309, FAX: 800-338-0309
SAFETY CONCEPTS, 5241 N 17th St., Ozark, MO 65721, 417-581-6199, FAX: 417-485-2156
SETON NAME PLATE CO., 20 Thompson Rd., Branford, CT 06405, 800-243-6624, FAX: 800-345-7819
STONEHOUSE SIGNS, INC., P.O. Box 546, Arvada, CO 80001, 800-525-0456, FAX: 800-255-0883

PLAQUES

BILL SIMS CO., 102 Lake Vista Dr., Chapin, SC 29063, 803-345-3606, FAX: 803-345-0315
CREATIVE AWARDS BY LANE, 1575 Elmhurst Rd., Elk Grove, IL 60009, 847-233-5666, FAX: 800-219-8777
CREATIVE OPTIONS USA, INC., 15 Lawrence Bell Dr., Amherst, NY 14221, 800-233-5666, FAX: 800-219-8777
D & L ASSOCIATES, 24795 Country Rd. #75, St. Cloud, MN 56301, 800-328-0307, FAX: 320-252-5504
IDESCO CORP., 37 W. 26th St., New York, NY 10010, 800-336-1383, FAX: 212-213-8078
LAB SAFETY SUPPLY, INC., 401 S. Wright Rd., Janesville, WI 53546, 800-356-0783, FAX: 800-543-9910
MPC PROMOTIONS, 2026 Shepardsville Rd., Louisville, KY 40218, 502-451-4900, 800-331-0989, FAX: 502-451-5075
NATIONAL SAFETY COUNCIL, 1121 Spring Lake Dr., Itasca, IL 60143, 630-285-1121, FAX: 630-285-1315
PEAVEY PERFORMANCE SYSTEMS, 14865 W. 105th St., Lenexa, KS 66215, 913-888-1095, FAX: 913-888-3898
POSITIVE IMPRESSIONS INC., 225 Westchester Ave., Port Chester, NY 10573, 914-937-8884, FAX: 914-937-6074
RAINBOW METALS INC., 17301 Beaton Rd. S.E., Monroe, WA 98272, 360-794-3691, FAX: 360-805-0815
SAFETY CONCEPTS, 5241 N. 17th St., Ozark, MO 65721, 417-581-6199, FAX: 485-2156
SETON NAME PLATE CO., 20 Thompson Rd., Branford, CT 06405, 800-243-6624, FAX: 800-345-7819
WILLIAMS JEWELRY & MFG. CO., 3152 Morris St. N., St. Petersburg, FL 33713, 813-823-7676, FAX: 813-822-2563

POCKET KNIVES

CONNEY SAFETY PRODUCTS, 3202 Latham Dr., Madison, WI 53713, 800-356-9100, FAX: 800-845-9095
CREATIVE AWARDS BY LANE, 1575 Elmhurst Rd., Elk Grove, IL 60009, 847-233-5666, FAX: 800-219-8777

CREATIVE OPTIONS USA, INC., 15 Lawrence Bell Dr., Amherst, NY 14221, 800-233-5666, FAX: 800-219-8777
D & L ASSOCIATES, 24795 Country Rd. #75, St Cloud, MN 56301, 800-328-0307, FAX: 320-252-5504
HAL REED CO., 12004 N.E. 172nd St., Kearney, MO 64060, 816-628-6722, FAX: 816-628-3295
LAB SAFETY SUPPLY, INC., 401 S. Wright Rd., Janesville, WI 53546, 800-356-0783, FAX: 800-543-9910
MPC PROMOTIONS, 2026 Shepardsville Rd., Louisville, KY 40218, 502-451-4900, 800-331-0989, FAX: 502-451-5075
POSITIVE IMPRESSIONS INC., 225 Westchester Ave., Port Chester, NY 10573, 914-937-8884, FAX: 914-937-6074
SERVICE MERCHANDISE CO., INC., 7100 Merchandise Dr, Brentwood, TN 37027, 800-367-7375, FAX: 615-660-7632
WILLIAMS JEWELRY & MFG. CO., 3152 Morris St. N., St. Petersburg, FL 33713, 813-823-7676, FAX: 813-822-2563

SPORTING GOODS

BILL SIMS CO., 102 Lake Vista Dr., Chapin, SC 29063, 803-345-3606, FAX: 803-345-0315
LL BEAN CORPORATE SALES, Casco St., Freeport, ME 04033, 207-865-4761, 800-832-1889, FAX: 207-797-6585
PEAVEY PERFORMANCE SYSTEMS, 14865 W. 105th St., Lenexa, KS 66215, 913-888-1095, FAX: 913-888-3898
WILLIAMS JEWELRY & MFG. CO., 3152 Morris St. N., St. Petersburg, FL 33713, 813-823-7676, FAX: 813-822-2563

TOOLS/HARDWARE

BILL SIMS CO., 102 Lake Vista Dr., Chapin, SC 29063, 803-345-3606, FAX: 803-345-0315
CONNEY SAFETY PRODUCTS, 3202 Latham Dr., Madison, WI 53713, 800-356-9100, FAX: 800-845-9095
CREATIVE AWARDS BY LANE, 1575 Elmhurst Rd., Elk Grove, IL 60009, 847-233-5666, FAX: 800-219-8777
CREATIVE OPTIONS USA, INC., 15 Lawrence Bell Dr., Amherst, NY 14221, 800-233-5666, FAX: 800-219-8777
D & L ASSOCIATES, 24795 Country Rd. #75, St. Cloud, MN 56301, 800-328-0307, FAX: 320-252-5504
MPC PROMOTIONS, 2026 Shepardsville Rd., Louisville, KY 40218, 502-451-4900, 800-331-0989, FAX: 502-451-5075
PEAVEY PERFORMANCE SYSTEMS, 14865 W. 105th St., Lenexa, KS 66215, 913-888-1095, FAX: 913-888-3898
WILLIAMS JEWELRY & MFG. Co., 3152 Morris St. N., St. Petersburg, FL 33713, 813-823-7676, FAX: 813-822-2563

TOTES AND BAGS

CREATIVE AWARDS BY LANE, 1575 Elmhurst Rd., Elk Grove, IL 60009, 847-233-5666, FAX: 800-219-8777
CREATIVE OPTIONS USA, INC., 15 Lawrence Bell Dr., Amherst, NY 14221, 800-233-5666, FAX: 800-219-8777
D & L ASSOCIATES, 24795 Country Rd. #75, St. Cloud, MN 56301, 800-328-0307, FAX: 320-252-5504
HAL REED CO., 12004 N.E. 172nd St., Kearney, MO 64060, 816-628-6722, FAX: 816-628-3295
LAB SAFETY SUPPLY, INC., 401 S. Wright Rd., Janesville, WI 53546, 800-356-0783, FAX: 800-543-9910
MPC PROMOTIONS, 2026 Shepardsville Rd., Louisville, KY 40218, 502-451-4900, 800-331-0989, FAX: 502-451-5075
POSITIVE IMPRESSIONS INC., 225 Westchester Ave., Port Chester, NY 10573, 914-937-8884, FAX: 914-937-6074
SERVICE MERCHANDISE CO., INC., 7100 Merchandise Dr., Brentwood, TN 37027, 800-367-7375, FAX: 615-660-7632
WILLIAMS JEWELRY & MFG. CO., 3152 Morris St. N., St. Petersburg, FL 33713, 813-823-7676, FAX: 813-822-2563

TRAVEL

BILL SIMS CO., 102 Lake Vista Dr., Chapin, SC 29063, 803-345-3606, FAX: 803-345-0315
PEAVEY PERFORMANCE SYSTEMS, 14865 W. 105th St., Lenexa, KS 66215, 913-888-1095, FAX: 913-888-3898

TROPHIES

BILL SIMS CO., 102 Lake Vista Dr., Chapin, SC 29063, 803-345-3606, FAX: 803-345-0315
CREATIVE AWARDS BY LANE, 1575 Elmhurst Rd., Elk Grove, IL 60009, 847-233-5666, FAX: 800-219-8777
CREATIVE OPTIONS USA, INC., 15 Lawrence Bell Dr., Amherst, NY 14221, 800-233-5666, FAX: 800-219-8777
D & L ASSOCIATES, 24795 Country Rd. #75, St. Cloud, MN 56301, 800-328-0307, FAX: 320-252-5504
MPC PROMOTIONS, 2026 Shepardsville Rd., Louisville, KY 40218, 502-451-4900, 800-331-0989, FAX: 502-451-5075
PEAVEY PERFORMANCE SYSTEMS, 14865 W. 105th St., Lenexa, KS 66215, 913-888-1095, FAX: 913-888-3898
POSITIVE IMPRESSIONS INC., 225 Westchester Ave., Port Chester, NY 10573, 914-937-8884, FAX: 914-937-6074
SAFETY CONCEPTS, 5241 N. 17th St., Ozark, MO 65721, 417-581-6199, FAX: 485-2156

WILLIAMS JEWELRY & MFG. CO., 3152 Morris St. N., St. Petersburg, FL 33713, 813-823-7676, FAX: 813-822-2563

UMBRELLAS

CREATIVE AWARDS BY LANE, 1575 Elmhurst Rd., Elk Grove, IL 60009, 847-233-5666, FAX: 800-219-8777
CREATIVE OPTIONS USA, INC., 15 Lawrence Bell Dr., Amherst, NY 14221, 800-233-5666, FAX: 800-219-8777
D & L ASSOCIATES, 24795 Country Rd. #75, St. Cloud, MN 56301, 800-328-0307, FAX: 320-252-5504
MPC PROMOTIONS, 2026 Shepardsville Rd., Louisville, KY 40218, 502-451-4900, 800-331-0989, FAX: 502-451-5075
POSITIVE IMPRESSIONS INC., 225 Westchester Ave., Port Chester, NY 10573, 914-937-8884, FAX: 914-937-6074
SERVICE MERCHANDISE CO., INC., 7100 Merchandise Dr., Brentwood, TN 37027, 800-367-7375, FAX: 615-660-7632
WILLIAMS JEWELRY & MFG. CO., 3152 Morris St. N., St. Petersburg, FL 33713, 813-823-7676, FAX: 813-822-2563

WRISTWATCHES

BENNETT BROTHERS, INC., 30 E. Adams St., Chicago, IL 60603, 312-621-1630, FAX: 312-621-1669
BILL SIMS CO., 102 Lake Vista Dr., Chapin, SC 29063, 803-345-3606, FAX: 803-345-0315
BULOVA CORP., One Bulova Ave., Woodside, NY 11377, 718-204-3331, FAX: 718-204-3546
CA SHORT CO., 4205 E. Dixon Blvd., Shelby, NC 28152, 704-482-9591, FAX: 704-484-3749
CREATIVE AWARDS BY LANE, 1575 Elmhurst Rd., Elk Grove, IL 60009, 847-233-5666, FAX: 800-219-8777
CREATIVE OPTIONS USA, INC., 15 Lawrence Bell Dr., Amherst, NY 14221, 800-233-5666, FAX: 800-219-8777
D & L ASSOCIATES, 24795 Country Rd. #75, St. Cloud, MN 56301, 800-328-0307, FAX: 320-252-5504
HAL REED CO., 12004 N.E. 172nd St., Kearney, MO 64060, 816-628-6722, FAX: 816-628-3295
LAB SAFETY SUPPLY, INC., 401 S. Wright Rd., Janesville, WI 53546, 800-356-0783, FAX: 800-543-9910
MPC PROMOTIONS, 2026 Shepardsville Rd., Louisville, KY 40218, 502-451-4900, 800-331-0989, FAX: 502-451-5075
NATIONAL SAFETY COUNCIL, 1121 Spring Lake Dr., Itasca, IL 60143, 630-285-1121, FAX: 630-285-1315
OMNITRAIN/SAFETY SHORTS, 2960 N. 23rd, La Porte, TX 77571, 800-458-2236, FAX: 713-470-8653

PEAVEY PERFORMANCE SYSTEMS, 14865 W. 105th St., Lenexa, KS 66215, 913-888-1095, FAX: 913-888-3898

POSITIVE IMPRESSIONS INC., 225 Westchester Ave., Port Chester, NY 10573, 914-937-8884, FAX: 914-937-6074

SAFETY SHORT PRODUCTIONS, 2960 N. 23rd St., La Porte, TX 77571, 800-458-2236, FAX: 713-470-8653

SERVICE MERCHANDISE CO., INC., 7100 Merchandise Dr., Brentwood, TN 37027, 800-367-7375, FAX: 615-660-7632

TIMEX CORP., Park Rd. Ext., P.O. Box 310, Middlebury, CT 06762, 203-573-5834, FAX: 203-573-5143

WILLIAMS JEWELRY & MFG. CO., 3152 Morris St. N., St. Petersburg, FL 33713, 813-823-7676, FAX: 813-822-2563

NOTES

PREFACE

[1] "What Do Employees Think About Your Safety Program," *Safety & Health*, November 1995, 42-43.

CHAPTER 1

[1] Ron Zemke and Dick Schaaf, *The Service Edge* (Markham, ON: Penguin Books, Canada, 1989), 72.

[2] Matt Green, "Panelists Split on Safety Incentives," *Canadian Occupational Safety*, May/June 1997, 4.

[3] *Collins English Dictionary* (Glasgow, Scotland: Harper Collins Publishers, 1994)

[4] Michael Hammer and Steven A. Stanton, *The Reengineering Revolution: A Handbook* (New York: Harper Collins, 1995), 129.

[5] Ibid.

[6] Ibid.

[7] "White Blasts Employers at Disability and Work Conferences," (St. John's, Newfoundland) *Evening Telegram*, 8 October 1996.

[8] *ENR*, 28 September 1989, 24.

[9] Robert Sass, "The Value of Safety Contests: A Point of View," *PAT Reporter*. June 1984, 13.

[10] K. Peter Richard, "The Westray Story: A Predictable Path to Disaster" Vol. 1 (Province of Nova Scotia, 1997), 75

[11] Ibid, 188.

[12] Judi Komaki, Kenneth Barwick and Lawrence Scott, "A Behavioral Approach to Occupational Safety: Pinpointing and Reinforcing Safe Performance in a Food Manufacturing Plant," *Journal of Applied Psychology* 63 (1978): 434-445.

[13] David K. Fox, B.L. Hopkins and W. Kent Anger, "The Long-Term Effects of a Token Economy on Safety Performance in Open Pit Mining," *Journal of Applied Behavior Analysis* 3 (Fall 1987): 223

[14] Gerald J.S. Wilde, *Target Risk* (Toronto: PDE Publications, 1994), 190-191.

[15] R. Bruce McAfee and Ashley R. Winn, "The Use of Incentives/Feedback to Enhance Work Place Safety: A Critique of the Literature," *Journal of Safety Research* 20 (1989): 9.

[16] Ibid,15-16.

[17] Alfie Kohn, *Punished by Rewards: The Trouble With Gold Stars, Incentive Plans, A's, Praise and Other Bribes* (New York: Houghton Mifflin, 1993), 38.

[18] Ibid, 3.

[19] Ibid, 41.

[20] Robert Howard, *Brave New Workplace—America's Corporate Utopias: How They Create Inequalities and Social Conflict in Our Working Lives* (New York: Viking Penguin, 1985), 120.

[21] John Micklethwait and Adrian Woolridge, *The Witch Doctors: Making Sense of the Management Gurus* (New York: Times Books/Random House, 1996), 206.

[22] Ibid, 209.

[23] Sass, 13.

[24] Green, 9.

[25] Mick Hans, "Safety for Hire: Can Companies Pay It Safe?" *Safety & Health*, May 1993, 45.

[26] Ibid.

[27] Kohn, 27.

[28] Ibid, 71.

[29] Ibid, 123.

[30] Peter R. Scholetes, "Reward and Incentive Programs are Ineffective—Even Harmful," *Small Business Forum*, Winter 1994-1995, 73.

CHAPTER 2

[1] Thomas A. Kochran, *Collective Bargaining and Industrial Relations: From Theory to Practice* (Chicago: Richard D. Irwin, 1980), 359.

[2] Dan Petersen, *Safety Management: A Human Approach* (New York: Aloray, 1988), 3-7.

[3] William C. Pope, *Managing for Performance Perfection: The Changing Emphasis* (Weaverville, NC: Bonnie Brae Publications, 1990), 129.

[4] W. Kip Viscusi, *Fatal Tradeoffs: Public and Private Responsibilities for Risk* (New York: Oxford University Press, 1992), 14.

[5] Ibid, 12.

[6] Douglas McGregor, *The Human Side of the Enterprise* (New York: McGraw Hill, 1960), 11.

[7] Ibid, 9.

[8] Aubrey C. Daniels, *Bringing Out The Best in People: How to Apply the Astonishing Power of Positive Reinforcement* (New York: McGraw Hill, 1994), 31

[9] Joan P. Klubnik, *Rewarding and Recognizing Employees: Ideas for Individuals, Teams and Managers* (Chicago: Richard D. Irwin, 1995), 105.

[1] Jane Coutts, "Workplace Stress More Prevalent Than Illness, Injury," *The* (Toronto) *Globe and Mail*, 18 April 1998.

[2] Charles Siler, "Do-It-Yourself Inspections, Scandinavian Style," *Safety & Health*, May 1994, 71.

[3] Viscusi, 1992.

[4] Ted Ferry, *Safety Program Administration for Engineers and Managers* (Chicago: Charles C. Thomas, 1984), 24-31.

[5] Petersen, 16.

[6] Ibid.

[7] Kohn, 3.

[8] Wayne Pardy, *Canadian Occupational Safety*, July/August 1993, 19-20.

[9] James M. Ham, "The Royal Commission on the Health & Safety of Workers In Mines" (Ontario, 1976), 119.

[10] Donald S. Barrie and Boyd C. Paulson, *Professional Construction Management* (New York: McGraw Hill, 1992), 402.

[11] Ibid, 403-404.

[12] Wayne Pardy, "Key Safety Issues for the Knowledge-Based 1990's," *Canadian Occupational Safety*, May/June 1993, 29.

[13] *Fortune*, 17 May 1993.

[14] *HR Canada*, August 1994.

[15] Richard Worzel, *The Next Twenty Years Of Your Life* (Toronto: Stoddart Publishing, 1997), 101-102.

[16] Pardy, "Key Safety Issues," 29.

[17] James R. Thomen, *Leadership in Safety Management* (New York: John Wiley & Sons, 1991), 23.

[18] Petersen, 113-114.

CHAPTER 4

[1] Kohn, 12-13.

[2] *Collins English Dictionary* (Glasgow, Scotland: Harper Collins Publishers, 1994).

[3] William E. Tarrants, *The Measurement of Safety Performance* (New York: Garland STMP Press, 1980), 243.

[4] Kohn, 187

[5] Edward E. Adams, "The Quality Revolution: Threat or Boon to Safety Professionals," *International Risk Control Review,* Fall 1988, 2.

[6] *Business Week*, 31 August 1992, 18.

[7] *The International Quality Study—Best Practices Report. An Analysis of Management Practices that Impact Performance.* (Ernst & Young/American Quality Foundation, 1992), 7.

[8] "Do the Right Thing: The Safety/Quality Relationship," *OHS Canada*, September/October 1991, 10.

[9] "Can Safety Be Too Much Fun," *Safety & Health,* September 1996, 27.

[10] Petersen, 124.

[11] E. Scott Geller, *The Psychology of Safety: How to Improve Behaviors and Attitudes on the Job* (Radnor: Chilton Book Company, 1996), 366.

[12] F. David Pierce, *Total Quality for Safety and Health Professionals* (Rockville, MD: Government Institutes, 1995), 163-164.

[13] Sandra O'Neal, "Study Shows Compensation Program Becoming More Strategic", *ACA News*, November/December 1996, 19-20.

[14] Michael J. Smith, Harvey H. Cohn, Alexander Cohn and Robert J. Cleveland, "Characteristics of Successful Safety Programs," *Journal of Safety Research*, 10, No. 1 (Spring 1978): 9-10.

[15] Wayne Pardy, "Winning and Losing: Ways to Reward Safety Performance," *Canadian Occupational Safety*, March/April 1994, 15.

[16] Ibid.

[17] Ibid.

[18] Nicholas A. Bartzis, SAFETY@LIST.UVM.EDU Internet response to questions on safety incentive approaches, March-April, 1998.

[19] Mark Hitz, SAFETY@LIST.UVM.EDU Internet response to questions on safety incentive approaches, March-April, 1998.

[20] Dave Adams, SAFETY@LIST.UVM.EDU Internet response to questions on safety incentive approaches, March-April, 1998.

[21] Mike Duram, SAFETY@LIST.UVM.EDU Internet response to questions on safety incentive approaches, March-April, 1998.

[22] Petersen, 126.

[23] Pardy, "Winning and Losing," 16-17.

CHAPTER 5

[1] E. Scott Geller, "Should You be Using Rewards?," *Industrial Safety & Hygiene News*, January 1995, 12.

[2] Ibid.

[3] Alfie Kohn, "Rewards Produce Temporary Compliance," *Small Business Forum*, 1994-1995, 67.

[4] Geller, "Should You be Using Rewards?," 12.

[5] Thomas R. Krause, John H. Hidley and Stanley J. Hodson, *The Behavior-Based Safety Process: Managing for an Injury Free Culture* (New York: Van Nostrand Reinhold, 1990), 19-20.

[6] Bob Filipczak, "Why No One Likes Your Incentive Program," *Training*, August 1993, 21.

[7] Krause, Hidley and Hodson, 45-46.

[8] "*Behavior-Based Safety: First Series Tool Box Booklets*" (Salt Lake City: Society for the Advancement of Safety & Health, 1997), 3.

[9] Catherine B. Kedjidjian, "What Do Employees Think About Your Safety Program," *Safety & Health*, November 1995, 42-43.

[10] McAfee and Winn, 15-17.

[11] Society for the Advancement of Safety and Health, 11-12.

[12] Geller, *The Psychology of Safety*, 178.

[13] Pardy, "Winning and Losing," 15.

[14] Ibid, 17.

[15] Michael R. Gilmore, "The Behavioral Approach," *OHS Canada,* November/December 1996, 26.

CHAPTER 6

[1] Aubrey Daniels, *Bringing Out the Best in People: How to Apply the Astonishing Power of Positive Reinforcement* (New York: McGraw-Hill, 1994), 4.

[2] Klubnik, 117.

[3] William E. Conway, *The Quality Secret: The Right Way To Manage* (Nashua: Conway Quality, 1992), 138.

[4] Krause, Hidley and Hodson, 38-40.

[5] Petersen, 27.

[6] Ibid, 106-107.

[7] Thomas J. Peters and Robert H. Waterman Jr. *In Search of Excellence: Lessons from America's Best-Run Companies* (New York: Harper & Row, 1982), 13-15.

[8] "Reich Reviews OSHA Reform," *Safety & Health*, July 1994, 28-30.

[9] "Interview with OSHA Policy Director John Moran," *Industrial Safety & Hygiene News*, March 1996,18.

[10] Ibid.

[11] Howard, 28-30.

[12] Ibid, 181.

[13] Micklethwait and Woolridge, 113-114.

[14] Bernard Wyscocki Jr., "The Danger of Stretching Too Far," *The Wall Street Journal*, August 1995.

[15] Hammer and Stanton, 128.

[16] Ibid, 129.

[17] Pierce, 14

[18] Alfie Kohn, *Punished By Rewards*, 41.

[19] "Positive Thinking: A Former OSHA Cop Becomes a Coach," *Industrial Safety & Hygiene News*, June 1994.

[20] *Safety & Health*, July 1995.

[21] Pierce, 83.

[22] *Fortune*, 18 September 1995.

[23] Petersen, 158.

[24] Judith Vogt and Kenneth Murrell, *Empowerment in Organizations: How to Spark Exceptional Performance* (University Associates, 1990).

CHAPTER 7

[1] Wayne Pardy, "A Measuring Stick for Safety Performance," *Canadian Occupational Safety*, March/April 1997, 36.

[2] "100 Performance Measures," *Industrial Safety & Hygiene News*, September 1995, 20.

CHAPTER 8

[1] Kevin A. Stewart and Wendy King, *The Role of Workplace Safety Incentives* (Hamilton, ON: Canadian Center for Occupational Health & Safety, 1991), 4-5.

[2] Mick Hans, "Safety for Hire—Can Companies Pay It Safe?," *Safety & Health*, May 1993, 47-48.

[3] SAFETY@LIST.UVM.EDU Internet responses to questions on the use of safety incentives, 1997-1998.

[4] Green, 4.

[5] Stewart & King, 6.

[6] Susan Harrelson, James Truss Co., SAFETY@LIST.UVM.EDU Internet response to questions on safety incentive approaches, March-April 1998.

[7] SAFETY@LIST.UVM.EDU Internet responses to questions on safety incentive approaches, March-April 1998.

[8] Bruce Brown, Safety Director, Atrium Companies, Inc., SAFETY@LIST.UVM.EDU Internet response to questions on safety incentive approaches, March-April 1998.

[9] Wilde, 194-199.

[10] Stewart & King, 8.

CHAPTER 9

[1] Wayne Pardy, *Canadian Occupational Safety*, July/August 1993.

[2] Ibid.

[3] Ilise Levy-Feitshans and Joseph E. Murphy, “Positive Incentives for Occupational Health: Addressing Workplace Violence” (Paper presented at the 25th International Congress on Occupational Health (ICOH), Stockholm, Sweden, 16 September 1996.)

[4] Ibid.

[5] Ibid.

[6] US Department of Labor, Occupational Safety & Health Administration, *So You Want to Apply to VPP? Here's How to Do It*! (Washington, DC: US GPO, 1997), 16.

[7] Levy-Feitshans and Murphy.

[8] Joseph Dear, 26 September 1995 Speech to Voluntary Protection Program Participants Association.

[9] US Department of Labor, Occupational Safety and Health Administration, "OSHA Resumes Traditional Enforcement Program: Partnership Remains on Hold," Press Release, 7 April 1998.

CHAPTER 10

[1] Jeremy Rifkin, *The End of Work: The Decline of the Global Labor Force at the Dawn of the Post-Market Era* (New York: G.P. Putnam's Sons, 1995, 182-183.

[2] Ibid, 189-190.

[3] Micklethwait and Woolridge, 206-207.

[4] Ibid, 16.

[5] Michael Nisbet, "Theory X, Like Tuberculosis, Is Making A Comeback," (Toronto) *Globe and M*ail, 16 February 1994.

[6] Klubnik, 133.

[7] "Will OSHA Regulate Incentive Programs?" *Occupational Hazards*, June 1998, 64.

[8] Ibid.

[9] Thomas B. Wilson, *Innovative Reward Systems for the Changing Workplace* (New York: McGraw Hill, 1995), 9.

[10] Ibid.

[11] William B. Abernathy, "Linking Performance Scorecards to Profit-Indexed Performance Pay," *ACA News*, April 1998, 23.

INDEX

A

ABC Analysis, 99
ACA News, 86, 259, 266
Abernathy, William B., 190, 266
Academic Skills, 69
Academics, 21, 166
Accident Causation, xii, 8, 52, 53, 57, 61, 87, 100
Accident Free, 4
Accident Investigation, 145
Accident Prevention, xiii, 25, 26, 45, 52, 70, 77, 109, 128, 132, 140, 156, 165, 192, 195, 196, 197, 199
Accident Proneness, 8
Accident-Based Safety Culture, xv
Accident-Free Incentives, 4
Achievement, v, xi, xv, xvi, 6, 8, 9, 59, 78, 87, 111, 139, 144, 150, 153, 157, 169, 170, 173, 175, 184, 186, 187, 189, 190, 192, 193, 195, 197, 200, 201
Achievement-Based Safety Culture, xv, xvi, 59, 169, 189, 190, 193
Activator-Behavior-Consequence, 86
Adams, Dave, 92-93, 259
Adams, Edward E., 79, 258
Administration Services, 235-236
Alarm Clocks, 240
American Quality Foundation, 82, 258
Anecdotal, 3
Anger, W. Kent, 14, 19, 254
Answering Machines, 240
Appraisal, 134
Appraisal Tools, 39
Asia, 87
Association of American Railroads, 117
Assumption about Human Behavior, 35
At-Risk Behavior, 111
Atrium Co., 163
Attitudes, xi, 8, 17, 23, 33, 69, 99, 100, 127, 157, 183
Audit, xvii, 9, 29, 34, 39, 41, 76, 91, 121, 138, 142, 144, 147, 151, 152, 153, 157, 158
Australia, xvii, 27
Award, 6
Award of Excellence, 201
Award of Merit, 201
Award Programs, xi, 84, 86, 165

B

Badges, 236
Bags, 250
Balance, 42, 70, 80, 110, 114, 121, 168, 187
Barrie, Donald S., 62-63, 257
Bartzis, Nicholas A., 90-91, 259
Barwick, Kenneth, 13, 254
Beck, Naula, 67
Behavior Manipulation, xiv, 17, 21
Behavior of Workers, xi
Behavioral Observation, 13
Behavior Based Safety, 260
Behavior-Based Safety, xii, 29, 35, 100, 106, 107, 108, 109, 111, 144
Behaviorist Tactics, 21
Behaviors, xi, xiv, 12, 13, 14, 23, 29, 43, 49, 63, 66, 67, 69, 75, 76, 85, 86, 88, 98, 99, 100, 104, 108, 109, 110, 111, 112, 113, 114, 117, 129, 138, 144, 149, 158, 165, 167, 183, 186, 189, 199
Belt Buckles, 236
Benchmark, 28, 127
Bennett Brothers Inc., 235, 237, 241, 242, 244, 251
Best Buy Co., 240, 242
Bill Sims Co., 237, 238, 242, 244, 245, 248, 249, 250
Blame-the-Victim Approach, 19
Blogg, John, 89-90, 113
Blueprint, xvi
Blunt Tool, 18
Bonus, 9, 10, 18, 50, 64, 101, 102, 103, 156, 157
Bracelets, 244-245

Brave New Workplace, 18, 122, 255
Bribing, 7, 17
Bringing Out the Best in People, 116, 256, 261
Broner Glove & Safety Co., 238
Brown, Bruce, 163, 264
Bulova Corp., 238, 244, 251
Burger King Corp., 235, 242
Business Week, 258

C

CA Short Co., 235, 237, 238, 239, 244, 251
CDs, 246
CEO, xvi, 5, 19, 20, 47, 48, 52, 61
Cameras, 237
Canada, xii, xvii, 4, 7, 8, 24, 27, 48, 49, 50, 51, 52, 65, 67, 69, 87, 90, 112, 129, 155, 159, 168, 174, 187, 253
Canada Health Monitor, 48, 49
Canadian Centre for Occupational Health and Safety, 155, 169, 263
Canadian Labor Congress, 7
Canadian Lake Carriers Association, 90
Canadian Occupational Safety, xvii, xxiii, 4, 65, 159, 174, 253, 257, 259, 263
Caps, 238-239
Carelessness, 8, 54, 127
Carrot and Stick, 5, 78, 97, 189
Cassettes, 246
Cause and Effect, 3, 116
Celebration, 6, 78
Central New York Council on Occupational Safety and Health, 102
Certif-A-Gift Co., 235, 242, 244
Challenge, xiii, xvi, 12, 24, 32, 38, 47, 56, 57, 60, 65, 70, 71, 74, 85, 95, 110, 131, 181, 185, 189, 190
China, 237
Claims Costs, 3
Class Issue, 20
Clemson University, 156
Cleveland, Robert J., 259
Clocks, 237-238
Clothing, 238-239
Cohen, A., 88-89
Cohn, Alexander, 259
Cohn, Harvey H., 259
Coincidental, 11
Collaboration, xv, 78, 128, 130
Collective Bargaining and Industrial Relations, 255
Collins English Dictionary, 253, 258
Command and Control, 4, 12, 21, 47, 49, 130
Commission Sales Schemes, 5
Committee of the Year Award, 198
Common Sense, 3, 16, 43, 57, 179
Compliance, xiii, xv, 12, 16, 17, 26, 27, 28, 29, 30, 31, 39, 53, 73, 76, 77, 97, 98, 106, 113, 120, 121, 128, 129, 131, 136, 146, 147, 148, 149, 151, 165, 173, 175, 177, 182, 187, 188, 196, 198
Comprehensive Occupational Safety and Health Reform Act, 120
Conference Board of Canada, 69
Conney Safety Products, 238, 239, 240, 241, 243, 249
Contest Contamination, 77
Contests, 7, 8, 22, 77, 78, 84, 89, 103
Contractor Safety Plan and Compliance, 149
Controversy, 6, 110
Conway, William E., 116, 261
Coolers, 239-240
Cooperative Compliance Program, 180
Coors Shenandoah Brewery, 129
Corev, S. Pressley, 129
Corporate Anorexia, 125
Corporate Culture, 115
Corporate Culture as Cults, 123
Cost Accounting, 64
Cost of Accidents, 24

Counterproductive, 4
Coutts, Jane, 256
Covey, Stephen R., 97
Creative Awards by Lane, 235, 236, 237, 238, 239, 240, 241, 242, 244, 245, 246, 247, 248, 249, 250, 251
Creative Business Solutions Inc, xvii, 145
Creative Options USA Inc., 235, 236, 237, 238, 239, 240, 241, 242, 244, 245, 246, 247, 248, 249, 250, 251
Critical Issues, 39
Criticism, 4, 94, 107, 167
Critique of Behavior-Based Safety, 106
Crosby, Phillip, 83
Crystal, 237
Cultural Changes, xiv
Cynicism, 22, 97, 115

D

D & L Associates, 235, 236, 237, 238, 239, 240, 241, 242, 244, 245, 246, 247, 248, 249, 250, 251
Daniels, Aubrey C., 116, 256, 261
Dear, Joseph, 179, 265
Debate, xi, xvi, 6, 10, 19, 23, 24, 29, 31, 50, 58, 61, 62, 73, 88, 92, 128, 156, 173, 185, 189
Deming Prize, 79
Deming, W. Edwards, 97
Dinners, 102
Dissatisfaction, xvi
Downsizing, 67, 70, 81, 82, 105, 125
Downstream, 116
Drucker, Peter F., 71
Due Diligence, 37, 42, 53, 135, 138, 146, 148, 195
Duram, Mike, 93-94, 260
Dysfunction, 8, 58, 63, 107, 116, 127

E

ENR, 7, 253
Economic and Industrial Democracy, 101
Economists, 174
Edmonds, Peter, 90
Electronics, 240
Emblem Pins, 236
Embroidered Patches, 246
Emergency Response, 148
Employee Ownership, 66, 101, 126, 134, 136
Employee Participation, 86, 109, 122, 182
Empowerment, 18, 47, 51, 66, 114, 126, 127, 128, 131, 133, 134, 135, 136, 166, 182
Empowerment in Organizations, 133
Enforcement, xiv, 26, 28, 65, 98, 121, 129, 131, 136, 152, 175, 177, 178, 179, 180
Ephemeral, 123
Ernst and Young, 82, 258
European Common Market, 27
Evening Telegram, 253
Executive Goals, 49

F

Fairfax, Rich, 188
Fatal Tradeoffs, 28, 54, 256
Fatality Risks, 30
Feeling/Emotional Recognition Options, 44
Ferry, Ted S., 54, 257
Financial Incentives, 5, 30, 174
Fire Blankets, 240-241
Fire Extinguishers, 240-241
Fire Protection Equipment, 240-241
First Aid Kits, 241
Fit-Rite Headwear, 238
Flashlights, 241-242
Fletcher Challenge Forests, 92
Flipczak, Bob, 260
Foods, 242
Fortune, 68, 130-131, 257, 263
Fox, David K., 14, 19, 254

Frequency, 9, 11, 12, 41, 89, 108, 109, 137, 148, 197
Fulton Industries, 241
Fundamentally Flawed, 57

G

Geller, E. Scott, 85-86, 97, 98, 259, 260, 261
General Duty Clause Violation, 188
General Electric, 106
Georgia Institute of Technology, 13
Germany, 21, 26
Gift Certificates, 242-243
Gilmore, Michael, 113, 261
Gimmicks, 97
Globe and Mail, 185, 256, 265
Government, xiii, xiv, xv, xvii, 4, 10, 17, 23, 24, 28, 30, 31, 47, 51, 58, 60, 64, 65, 68, 73, 120, 122, 125, 129, 135, 159, 173, 174, 175, 177, 187, 189, 190
Grandwell Industries Inc., 247
Great Britain. *See* United Kingdom.
Green, Mark, 253, 255, 263
Gurus, 56, 57, 72, 114, 135

H

HR Canada, 257
Hal Reed Co., 235, 237, 238, 239, 240, 241, 243, 244, 245, 246, 249, 250, 251
Ham Commission, 61, 257
Ham, James M., 61, 257
Hammer, Michael, 125, 253, 262
Hans, Mick, 255, 263
Hardware, 249
Harrelson, Susan, 264
Harvard University, 54
Hazard & Risk Analysis, 147
Healer Products Inc., 241
Health and Safety Program Award, 201
Heinrich, W.H., 25, 57, 61
Hidley, John H., 116, 260, 261
Highway Emergency Kits, 243-244
Hitz, Mark, 91-92, 259
Hodson, Stanley J., 116, 260, 261
Hopkins, B.L., 14, 19, 254
Housekeeping, 149
Howard, Robert, 18, 122-123, 254, 262

I

Ice Chests, 239-240
Idea Art, 243
Ideology, 18
Idesco Corp., 248
In Search of Excellence, 118, 262
Inbreeding, 65
Incentive(s), xi, xiv, xv, xvi, 3, 4, 5, 6, 7, 8, 9, 10, 11, 12, 15, 16, 18, 19, 20, 21, 22, 24, 25, 27, 28, 29, 30, 31, 32, 33, 34, 35, 37, 38, 39, 40, 42, 43, 45, 49, 50, 51, 52, 53, 55, 58, 63, 64, 65, 66, 67, 70, 75, 76, 78, 84, 85, 86, 87, 89, 90, 91, 92, 93, 97, 101, 102, 103, 104, 107, 108, 109, 110, 111, 113, 114, 121, 125, 139, 150, 155, 156, 157, 158, 159, 160, 162, 163, 165, 166, 168, 169, 170, 174, 175, 177, 179, 184, 185, 186, 187, 188, 189, 190, 194
Incentive Compensation, 5
Incentive Packages, 244
Incentive Products, 235-252
Incentive Rates, 36
Incentive Teams, 158
Individual Behavior, 87
Industrial Safety & Hygiene News, xxiv, 97, 121, 129, 150, 260, 262, 263
Injury Statistics, 4, 80, 99, 137, 139, 144
Injury-Based Culture, xvi
Innovations, 114
Innovative Reward Systems for the Changing Workplace, 188, 266
Innovative Safety Solution Award, 201

Inspection and Maintenance, 146
Internal Responsibility, 51, 52, 72, 187
International Conference on Occupational Health, 174, 264
International Labor Organization, 183
International Perspective, 87
International Risk Control Review, 258

J

JJ Keller & Associates, 243
Jackets, 238-239
James Truss Co., 264
Japan, 21, 79, 183
Jeffress, Charles N., 180
Jekyll and Hyde, 185
Jewelry, 244
John T. Ryan Trophy, 8, 9
Joint Health and Safety Committees, 145
Journal of Applied Behavior Analysis, 14, 254
Journal of Applied Psychology, 13, 254
Journal of Safety Research, 254, 259

K

Kansai Electric Power, 79-80
Karoshi, 183
Kedjidjian, Catherine B., 261
Key Chains, 245
King, Wendy, 155-156, 263, 264
Klubnik, Joan, 256, 261, 265
Kochran, Thomas A., 255
Kohn, Alfie, xxi, 16, 20, 21, 75, 78, 97-98, 128, 254, 255, 257, 258, 260, 262
Komaki, Judi, 13, 254
Krause, Thomas, 99, 107, 116, 260, 261

L

LL Bean Corporate Sales, 239, 243, 249
Lab Safety Supply, 236, 238, 239, 242, 245, 246, 247, 248, 249, 250, 251
Labelmaster, 236, 239, 243, 246, 247
Labor, ix, xi, xiii, xiv, xv, xvi, xvii, 4, 6, 10, 16, 24, 27, 33, 47, 58, 60, 61, 64, 65, 68, 69, 73, 75, 78, 112, 119, 122, 128, 130, 136, 176, 177, 182, 188, 189, 190, 198
Leadership, 43, 64, 65, 68, 69, 70, 71, 72, 73, 74, 120, 133, 178, 193, 195, 196, 198, 200, 201, 213
Leadership in Safety Management, 71, 257
Left Brain, 43
Lessen, Nancy, 100-107, 112
Levitt, Raymond, 63
Levy-Feitshans, Ilise, 174-177, 178, 264, 265
Liability Consequences, 62
Long-Term Objectives, 42, 99, 166
Louisiana State University, 93
Luggage, 245-246

M

ML Kishigo Mfg. Co., 239
MPC Promotions, 235, 236, 237, 238, 239, 240, 241, 242, 243, 244, 245, 246, 247, 248, 249, 250, 251
MSA Canada, 8
Macro Safety Issues, 72
Management Conceit, 22
Management Theorists, 184
Managerial Vigor, 166
Managing for Performance Perfection, 26, 256
Manipulation, xiv, 6, 17, 21, 107
Marshalls, 239, 243
Massachusetts Coalition for Occupational Health and Safety, 100
Massachusetts Institute of Technology, 183

McAfee, R. Bruce, 15, 110, 254, 261
McGregor, Douglas, 35-36, 256
Measurable Improvement, 77
Measures of Safety Performance, 137
Merit Pay, 22
Merit Program, 211
Micklethwait, John, 18, 123-124, 184, 254, 262, 265
Micro Safety Activities, 72
Milestone, 6, 59, 95
Mine Safety Appliances Co. of Canada, 8
Mines Accident Prevention Association of Manitoba, 89
Mobil Go Card, 235, 243
Morale, 80, 122, 129
Moran, John, 121, 262
Motivate, xi, xiii, 4, 6, 11, 19, 20, 21, 22, 32, 33, 55, 60, 67, 81, 84, 89, 91, 98, 99, 125, 157, 160, 175, 177, 178, 189
Multiple Measures, 80
Murphy, Joseph E., 174-177, 178, 264, 265
Murrell, Kenneth, 133, 263

N

NJ & Associates Inc., 235
National Institute of Occupational Safety and Health, 88
National Marker Co., 243, 248
National Safety Council, 19, 120, 235, 239, 245, 247, 248, 251
Necklaces, 244-245
Networking, 123
Newfoundland Power, xvii
Nisbet, Michael, 185, 265
North America, 8, 18, 29, 125, 127, 176, 198
Norway, 52

O

OH&S Committee, 73
OHS Canada, xvii, xxiii, 258, 261
OSF Inc, 90
OSHA, x, 24, 26, 30, 84, 106, 121, 129, 150, 153, 174, 175, 178, 179, 180, 188, 211, 212, 265
Obedience, xiii
Obfuscating Rhetoric, 54
Occupational Hazards, 188, 265
Occupational Safety and Health Act, 24, 26, 188
Omaha Steaks, 242, 243
Omnitrain/Safety Shorts, 238, 240, 245, 251
O'Neal, Sandra, 86, 259
Ontario Mining Association, 89, 113
Oregon Productivity Matrix, 91, 92
Organizational Freedom, 166
Ownership, 17, 47, 49, 50, 52, 66, 81, 92, 101, 126, 128, 131, 132, 134, 136

P

PAT Reporter, 253
PPM Safety Management Software, 145
Paradigm, 69, 126, 127
Pardy, Wayne G., xii, xxiii-xxiv, 65, 70, 174, 257, 259, 260, 261, 263, 264
Patches, 246
Paternalistic, 48, 55, 60, 66, 74
Paulson, Boyd C., 62-63, 257
Pay for Performance, 5
Peavey Performance Systems, 236, 237, 238, 239, 242, 243, 244, 245, 246, 247, 248, 249, 250, 252
Peer Pressure, 7, 102, 103, 157, 169
Penalties, 28, 97, 153, 175, 176
Pens, 246
Perceived Equity, 167
Perception Shift, 127
Perception Survey, 38, 39, 45, 151, 191
Performance Generators, 4
Performance Measurement, 139

Performance Measures, 41, 59, 139, 142, 143, 150, 190
Performance-Process-Measurement Safety Management Software, 145
Personal Management Skills, 69
Personal Protective Equipment, 147
Personal Stereos, 240
Peters, Tom, 118, 262
Petersen, Dan, 25, 26, 57, 59, 72-73, 85, 117, 132-133, 256, 257, 258, 260, 261, 263
Philosophical Arguments, 3
Pierce, F. David, xi-xiii, 86, 126-127, 130, 259, 262
Pins, 236, 244-245, 247
Pizza, 7
Plant Safety Scoreboards, 247-248
Plaques, 248
Pocket Knives, 248-249
Pope, William C., 26, 27, 256
Positive Attitude, 8
Positive Impressions Inc., 236, 237, 238, 239, 240, 241, 242, 243, 244, 245, 246, 247, 248, 249, 250, 251, 252
Positive Reinforcement, 13, 15, 90, 109, 149, 178
Power and Control, 18
Preferred Promotions Networks, 243
President's Award, 196
Pre-Work Planning (Tool Box Talks), 149
Principle of Congruence, 133
Principle of Coordination, 134
Principle of Excellence, 133
Principle of Integration, 134
Principle of Interdependence, 133
Principle of Investment, 134
Principle of Process and Direction, 134
Principles, xiii, 3, 10, 26, 34, 47, 53, 60, 75, 77, 83, 85, 87, 88, 112, 133, 134, 138, 193, 195
Prizes, 6
Proaction Communication Services Corp., 235
Production Bonus System, 9
Production Quota, 105
Productivity, xiv, 10, 72, 125, 129, 182, 183
Professional Construction Management, 62, 257
Promotions Modify Behavior, 91
Punished, 31
Punished by Rewards, 75, 97, 254, 262
Punishment as a Form of Deterrence, 176
Punishments, 16, 20

Q

Quality, 129
Quality is Free, 83

R

Radios, 240
Rainbow Metals Inc., 236, 247, 248
Raytheon TI Systems, 91
Recognition, xi, xii, xv, xvi, 3, 4, 5, 9, 12, 16, 20, 21, 24, 25, 29, 31, 33, 34, 35, 37, 38, 39, 40, 43, 44, 45, 51, 52, 58, 59, 60, 65, 66, 67, 70, 78, 81, 84, 87, 88, 89, 90, 92, 94, 97, 99, 109, 110, 111, 113, 114, 116, 127, 139, 140, 141, 144, 145, 150, 161, 165, 169, 170, 174, 175, 184, 186, 187, 188, 189, 192, 195, 196, 198, 200, 211
Reich, Robert B., 120, 262
Reinforcement, 13, 15, 16, 17, 66, 75, 90, 109, 149, 153, 167, 178, 186, 189, 196
Responsibility for Safety, 50
Result Measures, 137
Reward People for Safety, xi
Reward System, 3
Reward Systems, 86
Rewarded, xvi, 3, 15, 18, 21, 78, 86, 90, 94, 110, 111, 156, 167, 175, 187
Rewarding and Recognizing Employees, 256

Rewarding Safe Behaviors, 104
Rhetoric, 18, 29, 47, 54, 74, 115, 127, 135
Richard, K. Peter, 10, 254
Rifkin, Jeremy, 182, 265
Right Brain, 43
Rings, 244-245
Risk Management, xvii, 9, 27, 39, 55, 100, 144, 147
Root Causes, 6, 162

S

Saf-T-Guard International Inc., 241
Safe Behavior Observation System, 149
Safety & Health, 19, 26, 84, 86, 88, 102, 106, 120, 126, 129, 130, 179, 188, 253, 255, 256, 258, 261, 262, 263
Safety Audits, 76, 147
Safety Concepts, 235, 239, 244, 248, 250
Safety Culture, xv, xvi, 9, 31, 46, 57, 59, 100, 115, 118, 119, 121, 134, 139, 166, 168, 169, 173, 185, 189, 190, 193
Safety Incentive, xi, xii, xiv, xcv, xvi, 3, 4, 6, 7, 9, 10, 11, 12, 15, 16, 19, 20, 21, 30, 31, 32, 33, 34, 35, 37, 38, 40, 43, 45, 49, 52, 55, 58, 63, 65, 66, 67, 84, 85, 86, 87, 89, 92, 93, 97, 101, 102, 103, 104, 110, 111, 125, 150, 155, 156, 157, 158, 159, 160, 162, 163, 165, 166, 168, 169, 170, 184, 186, 187, 189
Safety Incentive Guide, 169
Safety Leadership Recognition Awards, 197
Safety Legislation, xii, 27, 28, 31, 50, 120, 176
SAFETY@LIST.UVM.EDU, 157 259, 260, 263, 264
Safety Management, 57, 72-73, 85, 117, 256
Safety Management Strategy, 34, 37, 64, 100
Safety Maturity Level, 38, 42
Safety Meetings, 146
Safety Objective Setting, 145
Safety Performance Measurement, 142
Safety Performance Measurement Software, 145
Safety Performance Solutions, 113
Safety Records, 64
Safety Regulations, 27, 29, 114
Safety/Quality Relationship, 80, 83
Safety Short Productions, 240, 252
Sass, Robert, 8, 18, 19, 253, 255
Scandinavia, 51, 112
Schaaf, Dick, 4, 253
Schizophrenia, xiii
Scholtes, Peter R., 22, 255
Science, 19, 27, 53, 54, 61, 67
Scientific Management, 19, 122, 184
Scott, Lawrence, 13, 254
Sears Auto Centers, 107
Seduction, 20
Self-Regulation, 51
Senior Management, xv, xvi, 15, 32, 37, 38, 43, 63, 99, 133, 162
Service Merchandise, 236, 237, 238, 240, 241, 242, 243, 244, 245, 246, 247, 249, 250, 251, 252
Seton Name Plate Inc., 236, 247, 248
Severity, 9, 11, 12, 41, 89, 102, 108, 109, 137, 197
Shared Beliefs, 47
Shifting Gears, 67
Shirts, 238-239
Simoneau, Barrie, 89, 95
Siler, Charles, 256
Skinner, B.F., 75
Slogans, 13, 51, 53, 54, 55, 56, 57, 77, 185
Small Business Forum, 22, 255, 260
Smith, Michael J., 88-89, 259
Smoke Detectors, 240-241

Society for the Advancement of Safety and Health, xiii, 260, 261
Software Tools, 181
Sonoco Products, 129
Sony Music Special Products, 246
Spiegel Inc., 239, 243
Sporting Goods, 249
Standards, xii, xiii, xv, 17, 27, 28, 29, 30, 60, 62, 72, 73, 77, 78, 88, 121, 132, 141, 143, 144, 145, 147, 148, 149, 174, 179, 186, 196
Stanton, Steven A., 125, 126, 253, 262
Star Program, 211
Statistically Reliable, 137
Stereos, 240
Stewart, Kevin A., 155-156, 263, 264
Stock Yards Chicago, 242, 243
Stonehouse Signs Inc., 248
Stored Value Marketing, 243
Strategic, xiv, 3, 12, 31, 35, 36, 37, 38, 39, 40, 41, 42, 43, 45, 50, 52, 55, 60, 65, 82, 86, 133, 134, 135, 139, 140, 141, 142, 150, 186, 187
Strategic Planning, 41
Stress, 48, 67, 183
Success Stories, 12, 135, 173
Sunoco Products, 129
Supervisory Development and Safety Management, 146
Survivor Syndrome, 125
Sweatshirts, 238-239
Symptomatic, 63
Syncrude, 113

T

TJ Maxx, 239, 243
TJX Corp. Incentives, 239, 243
TQM. *See* Total Quality Management.
Tapes, 246
Target Risk, 167, 254
Tarrants, William E., 77, 258
Taylor, Frederick, 18
Team-Based Awards, 102
Team-Based Safety, 66
Teamwork, xvi, 41, 80, 128, 129, 130, 182
Teamwork Skills, 69
Telephones, 240
The Behavior-Based Safety Process, 99, 116, 260
The End of Work, 182, 265
The Human Side of the Enterprise, 35, 256
The International Quality Study, 258
The Measurement of Safety Performance, 77, 258
The Next Twenty Years of Your Life, 70, 257
The Psychology of Safety, 85, 259, 261
The Quality Revolution, 79, 258
The Quality Secret, 116, 261
The Reengineering Revolution, 5, 125, 253
The Role of Workplace Safety Incentives, 160, 263
The Service Edge, 4, 253
The Team Handbook, 22
The Value of Safety Contests, 8
The Witch Doctors, 18, 123, 254
Theoretical Evidence, 3
Theory X, 184, 185, 186
Theory Y, 184, 185
Therapeutic, 123
Thinking (Less Personal) Recognition Options, 44
Third World, 27
Thomen, James R., 71, 257
Threatening, 17, 21
Tie Tacks, 244-245
Timex Corp., 252
Token, xiv, 5, 14, 16, 19, 20, 67, 198
Token Compliance, 131
Tolls, 249
Total Quality for Safety and Health Professionals, 86, 126-127, 130, 259
Total Quality Management, xiii, xiv, 26, 63, 77, 80, 82, 83, 109, 114, 128, 182

Total Safety Management, 10, 58, 64, 77, 187, 194
Totes, 250-251
Towers Perrin, 86
Toys "R" Us Incentive Sales, 236, 243
Tradeoffs, 174
Trading Stamps, 14
Training, 260
Travel, 250
Trophies, 250-251
T-Shirts, 238-239

U

US Department of Labor, 178, 211
Umbrellas, 251
Underreporting of Injuries, 103
Unions, 23, 28, 58, 61, 64, 122, 125
United Kingdom, 26, 51, 183
United States, 21, 27, 28, 90, 183
University of Colorado, 101
University of Puget Sound, 101
University of Saskatchewan, 18
University of Southern California, 54
Unsafe Act-Unsafe Condition Variable, 61
Unsafe Behavior, xii, 13, 57, 59, 104, 108, 112, 138, 167
Unsafe Conditions, 8, 23, 61, 62, 63
Upstream, 117

V

Values, 43
Videocassette Recorders, 240
Viscusi, W. Kip, 28, 30-31, 54, 174, 256
Vogt, Judith, 133, 263
Voluntary Protection Program, 175, 178, 179, 211, 265
Volunteer Appreciation Award, 201
VPP. *See* Voluntary Protection Program.

W

Wage Compensation, 30
Wall Street Journal, 125, 262
Waterman, Robert, 118, 262
Wesley, John, 129
Westray, ii, 8, 9, 10, 50
Westray Mine Public Inquiry Report, 8, 254
White, Bob, 7, 253
Wilbert Co., 19
Wilde, Gerald J.S., 167, 254, 264
Wilkerson, Bruce, 163
Williams Jewelry & Mfg. Co., 236, 237, 238, 239, 240, 241, 242, 244, 245, 246, 247, 248, 249, 250, 251, 252
Williams, Jim, 113
Wilson, Thomas B., 188-189, 266
Winn, Ashley R., 15, 110, 254, 261
Woolridge, Adrian, 18, 123-124, 184, 254, 262, 265
Workplace Health, Safety and Compensation Commission of Newfoundland and Labrador, 203
Workplace Violence, 264
Worker Motivation, 22
Workers' Compensation, 150
Worzel, Richard, 70, 257
Wristwatches, 251-252
Wyscocki, Bernard, Jr., 125, 262

Z

Zeckhauser, Richard, 53-54
Zemke, Ron, 4, 253
Zero Injuries, 92

NOTES

NOTES

You Are Invited...

to join the

Society for the Advancement of Safety and Health

The Society for the Advancement of Safety and Health was formed in 1997 to promote innovative concepts and expand the horizons of workplace safety and health. SASH provides members with a variety of safety and health resources, including two newsletters, a series of "Tool Box for Success" booklets, Proficiency Certifications in safety, environment and management, and discounts on safety literature and training course registration.

Individual and company memberships are available.

RSVP

For more information, visit our Web site at
http://www.sash-org.com
or call, write or fax:

You Are Invited...

to join the

Society for the Advancement of Safety and Health

The Society for the Advancement of Safety and Health was formed in 1997 to promote innovative concepts and expand the horizons of workplace safety and health. SASH provides members with a variety of safety and health resources, including two newsletters, a series of "Tool Box for Success" booklets, Proficiency Certifications in safety, environment and management, and discounts on safety literature and training course registration.

Individual and company memberships are available.

RSVP

For more information, visit our Web site at
http://www.sash-org.com
or call, write or fax: